Texts in Applied Mathematics 47

Editors
J.E. Marsden
L. Sirovich
S.S. Antman

Advisors
G. Iooss
P. Holmes
D. Barkley
M. Dellnitz
P. Newton

Springer

New York
Berlin
Heidelberg
Hong Kong
London
Milan
Paris
Tokyo

Texts in Applied Mathematics

(continued after index)

Hilary Ockendon   John R. Ockendon

Waves and Compressible Flow

With 60 Figures

Springer

Hilary Ockendon
Oxford Centre for Industrial and
   Applied Mathematics
24–29 St. Giles
Oxford OX1 3LB
UK
ockendon@maths.ox.ac.uk

John R. Ockendon
Oxford Centre for Industrial and
   Applied Mathematics
24–29 St. Giles
Oxford OX1 3LB
UK
ock@maths.ox.ac.uk

Series Editors
J.E. Marsden
Control and Dynamical Systems, 107–81
California Institute of Technology
Pasadena, CA 91125
USA
marsden@cds.caltech.edu

L. Sirovich
Division of Applied Mathematics
Brown University
Providence, RI 02912
USA
chico@camelot.mssm.edu

S.S. Antman
Department of Mathematics
and
Institute of Physical Science
   and Technology
University of Maryland
College Park, MD 20742-4015
USA
ssa@math.umd.edu

Mathematics Subject Classification (2000): 76-02, 76Nxx, 76Bxx

Library of Congress Cataloging-in-Publication Data
Ockendon, Hilary.
   Waves and compressible flow / Hilary Ockendon, John R. Ockendon.
      p. cm. — (Texts in applied mathematics ; v. 47)
   Includes bibliographical references and index.

   1. Wave motion, Theory of.   2. Fluid dynamics.   3. Compressibility.   I. Title.   II. Texts in
   applied mathematics ; 47.
QA927.O25 2003
532′.0535—dc21   2003054314

ISBN   978-1-4419-2335-6   e-ISBN   978-0-387-21802-1   Printed on acid-free paper.

Printed in the United States of America.   (BPR/MVY)

9 8 7 6 5 4 3 2 1

Springer-Verlag is a part of *Springer Science+Business Media*

springeronline.com

Series Preface

Mathematics is playing an ever more important role in the physical and biological sciences, provoking a blurring of boundaries between scientific disciplines and a resurgence of interest in the modern as well as the classical techniques of applied mathematics. This renewal of interest, both in research and teaching, has led to the establishment of the series Texts in Applied Mathematics (TAM).

The development of new courses is a natural consequence of a high level of excitement on the research frontier as newer techniques, such as numerical and symbolic computer systems, dynamical systems, and chaos, mix with and reinforce the traditional methods of applied mathematics. Thus, the purpose of this textbook series is to meet the current and future needs of these advances and to encourage the teaching of new courses.

TAM will publish textbooks suitable for use in advanced undergraduate and beginning graduate courses, and will complement the Applied Mathematical Sciences (AMS) series, which will focus on advanced textbooks and research-level monographs.

Pasadena, California   J.E. Marsden
Providence, Rhode Island   L. Sirovich
College Park, Maryland   S.S. Antman

Contents

The starred sections are self-contained and may be omitted at a first reading.

1

Introduction

These lecture notes have grown out of a course that was conceived in Oxford in the 1960s, was modified in the 1970s and formed the basis for *Inviscid Fluid Flows* by Ockendon and Tayler which was published in 1983 [1]. This monograph has now been retitled and rewritten to reflect scientific development in the 1990s.

The cold war was at its height when Alan Tayler gave his first course on Compressible Flow in the early 1960s. Naturally, his material emphasized aeronautics, which was soon to be encompassed by aerospace engineering, and it concerned flows ranging from small-amplitude acoustics to large-amplitude nuclear explosions. The area was technologically glamorous because it described how only mathematics could give a proper understanding of the design of supersonic aircraft and missiles. It was also mathematically glamorous because the prevalence of "shock waves" in the physically relevant solutions of the equation of compressible flow led many students into a completely new appreciation of the theory of partial differential equations. Suddenly, there was the challenge to find not only non-differentiable but also genuinely discontinuous solutions of the equations and the simultaneous problem of locating the discontinuity. This led to enormous theoretical developments in the theory of *weak solutions* of differential equations and, more generally, to the whole theory of *moving boundary problems*.

It has been the even more dramatic developments that have occurred recently in all branches of applied science that have made the scope of this book so much broader than that of its predecessor. In particular, three recent "revolutions" have changed the mathematical aspects of compressible flow and, more generally, of wave motion.

First, the computer revolution has completely altered the way mathematicians need to think about systems of partial differential equations. Gone is the need for academic "exact solutions", or for *ad hoc* approximate solutions. In their place, mathematics now has to provide all-important guides to well-posedness and to systematic perturbation theories that can provide quality control for scientific computation, especially in parameter regimes that are awkward to analyze numerically. Of course, physically relevant exact solutions are still invaluable for the insight they give, but more and more they are used as checks on computer output.

Second, the communications revolution has immeasurably increased the demand for understanding electromagnetic waves in situations that were no more than science fiction in the 1960s. An applied mathematician working in the real world may now have to have a good theoretical understanding of the working of optical fibers, radio waves in cluttered environments, and the waves generated by electronic components. All of these phenomena are governed by wave equations not too dissimilar from those arising in gasdynamics, but in configurations that call out for completely new solution methods.

Third, the environmental revolution has presented the whole community with a host of new problems associated with wave propagation in the atmosphere, in the oceans, and in the interior of the earth. The models describing these waves are often much more complicated than those from compressible flow, involving far more mechanisms and, especially, wildly disparate time and length scales. Nonetheless, we will see that, in many situations, these models are still susceptible to the traditional methodologies devised for treating gasdynamics. We should also mention the importance of waves in solids in connection with modern developments in materials science and non-destructive testing.

Even after these upheavals, it remains the authors' abiding belief that fluid mechanics provides the best possible vehicle for anyone wishing to learn applied mathematical methodology, simply because the phenomena are at once so familiar and so fascinatingly complex. Indeed, the mathematical study of these phenomena has led to some of the most dramatic new ideas in the theory of partial differential equations as well as profound scientific insights that have affected much of the modern theoretical framework in which we understand the world around us.

In the light of these developments, the lecture course on which this book is based has undergone an organic transformation in order to provide students with a basis for understanding the wide range of wave phenomena with which any applied mathematician may now be confronted. Hence, this monograph reflects a shift in emphasis to one in which gasdynamics is seen as a paradigm for wave propagation more generally and in which the associated mathematics is presented in a way that facilitates its wider use. Although compressible flow remains the main focus of the book, and we still derive the equations of compressible flow in some detail, we will also show how wave phenomena in electromagnetism and solid mechanics can be treated using similar mathematical methods. We cannot give a comprehensive account of models for these other kinds of waves nor can we, in the space available, even start to describe the burgeoning area of mechanical and chemical wave propagation in biological systems. However, we will revisit their omission in the Epilogue and provide some references to relevent texts at the same level as this one.

The layout of the book is as follows. We begin in Chapter 2 with a derivation of the equations of compressible flow that is as simple as possible while still being self-contained. The only required physical background is a belief in the ideas of conservation of mass, momentum, and energy together with the associated elementary thermodynamics. Then, in Chapter 3, we immediately distill the simplest wave motion model to emerge from the general equations of gasdynamics, namely the model for acoustics. This will be applied not only to sound propagation and to some theories of flight but, before that, we will present several other models for linear wave propagation that are relevant to the fields of application listed above. Except for the case of surface gravity waves, these will take the form of linear hyperbolic partial differential equations for which, thankfully, there is a fairly well-developed body of knowledge, even at the undergraduate level. We will recall some of the more important exact solutions in Chapter 4 and the phenomena that they reveal, especially that of dispersion. Then, we will look more generally at waves that have a purely harmonic time-dependence, sometimes called monochromatic waves or waves in the frequency domain. This assumption frequently reduces the linear models of Chapter 3 to elliptic partial differential equations, which are also well studied at the undergraduate level, but the questions that need to be answered are often very different from those traditionally associated with elliptic equations. Following on from this, we look at high-frequency (which often means short-wavelength) approximations in frequency domain models. This leads us to the ever-more-important "ray theory" approach to wave propagation which, as we will see, opens up fascinating new mathematical challenges and analogies in subjects ranging from quantum mechanics to celestial mechanics.

In Chapter 5, we return to our generic theme of compressible flow with a review of the little that is known about nonlinear solutions, followed by the similarly meager theory for nonlinear surface gravity waves. Finally, in Chapter 6, we will present a theory that allows us to consider shock waves and the sound barrier and helps us to understand several other interesting nonlinear phenomena such as laminar and turbulent nozzle flows, detonations, and transonic and hypersonic flows.

This book is written, as was its predecessor, at a level that *assumes* that the reader already has some familiarity with basic fluid dynamics modeling, especially the use of the convective derivative and the basis of the Euler equations for incompressible flow. A knowledge of asymptotic analysis up to Laplace's method and the method of stationary phase is also helpful; we do not have space to give ab initio accounts of these methods, which underpin the mathematics of group velocity and ray theory, but we do give a brief recapitulation and references to texts where the reader can find all the details.

The starred sections are self-contained and describe more advanced topics which can be omitted at a first reading. The exercises are an integral part of the book; those marked R are "recommended" as containing basic material, whereas the starred ones are harder or refer to the work in starred sections.

Both authors acknowledge their great debt to their guide and mentor Alan Tayler; it will be apparent to all who knew him that this book is part of his rich legacy to applied mechanics. Also we would like to record our special thanks to Brenda Willoughby for her invaluable assistance with the preparation of this book and to Carina Edwards whose suggestions have greatly enhanced its presentation.

2

The Equations of Inviscid Compressible Flow

In this chapter, we will derive the equations of inviscid compressible flow of a perfect gas. We will do this by making the traditional assumption that we are working on length scales for which it is reasonable to model the gas as a *continuum*; that is to say, it can be described by variables that are smoothly defined[1] almost everywhere. This means that the gas is infinitely divisible into smaller and smaller *fluid elements* or *fluid particles* and we will see that it will help our understanding to relate these particles to the "particles" of classical mechanics.

This approach will, of course, become physically inaccurate at small enough scales because all matter is composed of molecules, atoms, and sub-atomic particles. This is particularly evident for gases especially when they are in a rarified state as, for example, is the case in the upper atmosphere. In order to treat such gases when the mean free path of the molecules is large enough to be comparable with the other length scales of interest (such as the size of a space vehicle), it is necessary to resort to the ideas of statistical mechanics. As described in Chapman and Cowling [2], this leads to the well-developed, but much more difficult *kinetic theory of gases* and, fortunately, when the limit of this theory is taken, on a scale which is much greater than a mean free path, the equations which we derive in this chapter can be retrieved.

2.1 The Field Equations

With the continuum approach, the state of a gas may be described in terms of its velocity $\mathbf{u}$, pressure p, density ρ, and absolute temperature T. If the independent variables are $\mathbf{x}$ and t, where $\mathbf{x}$ is a three-dimensional vector with components either (x, y, z) or (x_1, x_2, x_3) referred to inertial cartesian axes and t is time, then we have an *Eulerian* description of the flow. An alternative

---

[1] We hope the reader will not be deterred by such imprecision, which is necessary to keep applied mathematics texts reasonably concise.

description, in which attention is focused on a fluid particle, is obtained by using $\mathbf{a}$, and t as independent variables, where $\mathbf{a}$ is the initial position of the particle. This is a *Lagrangian* description. A particle path $\mathbf{x} = \mathbf{x}(\mathbf{a}, t)$ is obtained by integrating $\dot{\mathbf{x}} = \mathbf{u}$ with $\mathbf{x} = \mathbf{a}$ at $t = 0$, where the dot denotes differentiation with respect to t keeping $\mathbf{a}$ fixed, and this relation may be used to change from Eulerian to Lagrangian variables. The two descriptions are equivalent, but for most problems, the Eulerian variables are found to be more useful.[2]

It is important to distinguish between differentiation "following a fluid particle," which is denoted by d/dt, and differentiation at a fixed point, denoted by $\partial/\partial t$. If $f(\mathbf{x}, t)$ is any differentiable function of the Eulerian variables $\mathbf{x}$ and t, then

$$\frac{df}{dt} = \frac{\partial f}{\partial t} + (\mathbf{u} \cdot \nabla)f, \qquad (2.1)$$

where ∇ is the gradient operator with respect to the $\mathbf{x}$ components. The derivative df/dt is called the *convective derivative* and the term $(\mathbf{u} \cdot \nabla)f$ is the convective term which takes account of the motion of the fluid.

We have already assumed that the fluid is a continuum and this implies that the transformation from $\mathbf{a}$ to $\mathbf{x}$ is, in general, a continuous mapping which is one-to-one and has an inverse. We will also restrict attention to flows for which this mapping is continuously differentiable almost everywhere. The Jacobian of the transformation, $J(\mathbf{x}, t) = \partial(x_1, x_2, x_3)/\partial(a_1, a_2, a_3)$, represents the physical dilatation of a small element. In order to understand the evolution of a fluid flow, it will be helpful to work out how J changes following the fluid. Since the transformation from $\mathbf{a}$ to $\mathbf{x}$ is invertible and continuous, J will be bounded and non-zero and its convective derivative will be

$$\frac{dJ}{dt} = \frac{\partial(\dot{x}_1, x_2, x_3)}{\partial(a_1, a_2, a_3)} + \frac{\partial(x_1, \dot{x}_2, x_3)}{\partial(a_1, a_2, a_3)} + \frac{\partial(x_1, x_2, \dot{x}_3)}{\partial(a_1, a_2, a_3)}$$

$$= \frac{\partial(u_1, x_2, x_3)}{\partial(a_1, a_2, a_3)} + \frac{\partial(x_1, u_2, x_3)}{\partial(a_1, a_2, a_3)} + \frac{\partial(x_1, x_2, u_3)}{\partial(a_1, a_2, a_3)}.$$

Writing out the first term, we see that

$$\frac{\partial(u_1, x_2, x_3)}{\partial(a_1, a_2, a_3)} = \begin{vmatrix} \dfrac{\partial u_1}{\partial a_1} & \dfrac{\partial u_1}{\partial a_2} & \dfrac{\partial u_1}{\partial a_3} \\ \dfrac{\partial x_2}{\partial a_1} & \dfrac{\partial x_2}{\partial a_2} & \dfrac{\partial x_2}{\partial a_3} \\ \dfrac{\partial x_3}{\partial a_1} & \dfrac{\partial x_3}{\partial a_2} & \dfrac{\partial x_3}{\partial a_3} \end{vmatrix}.$$

---

[2] We make this remark in the context of understanding the *mathematical* basis of models for compressible flow. For computational fluid dynamics, particle-tracking methods are often more appropriate than discretizations based on Eulerian variables.

However,

$$\frac{\partial u_1}{\partial a_i} = \frac{\partial u_1}{\partial x_1}\frac{\partial x_1}{\partial a_i} + \frac{\partial u_1}{\partial x_2}\frac{\partial x_2}{\partial a_i} + \frac{\partial u_1}{\partial x_3}\frac{\partial x_3}{\partial a_i},$$

and so, using the properties of determinants, we obtain

$$\frac{\partial(u_1, x_2, x_3)}{\partial(a_1, a_2, a_3)} = J\frac{\partial u_1}{\partial x_1}.$$

The other two terms can be treated similarly and so

$$\frac{dJ}{dt} = J\nabla \cdot \mathbf{u}. \tag{2.2}$$

We can now consider the rate of change of any property, such as the total mass or momentum, in a *material volume* $V(t)$, which is defined as a volume which contains the same fluid particles at all times. We find that

$$\frac{d}{dt}\left[\int_{V(t)} F(\mathbf{x}, t)\, dV(\mathbf{x})\right] = \frac{d}{dt}\left[\int_{V(0)} F(\mathbf{x}(\mathbf{a}, t), t)J\, dV(\mathbf{a})\right]$$

$$= \int_{V(0)} \frac{d}{dt}[F(\mathbf{x}(\mathbf{a}, t), t)J]dV(\mathbf{a})$$

$$= \int_{V(0)} \left(\frac{dF}{dt}J + FJ\nabla \cdot \mathbf{u}\right) dV(\mathbf{a}) \quad \text{(on using (2.2))}$$

$$= \int_{V(t)} \left(\frac{dF}{dt} + F\nabla \cdot \mathbf{u}\right) dV(\mathbf{x}). \tag{2.3}$$

This formula for differentiating over a volume which is "moving with the fluid" is called the *transport theorem*. Using (2.1) and denoting the outward normal to $\partial V(t)$ by $\mathbf{n}$, we can rewrite (2.3) as

$$\frac{d}{dt}\left[\int_{V(t)} F\, dV\right] = \int_{V(t)} \left(\frac{\partial F}{\partial t} + \nabla \cdot (F\mathbf{u})\right) dV \tag{2.4}$$

$$= \int_{V(t)} \frac{\partial F}{\partial t}dV + \int_{\partial V(t)} F\mathbf{u} \cdot \mathbf{n}\, dS, \tag{2.5}$$

on using the divergence theorem. Thus, from (2.5), the derivative can be interpreted as the sum of the term $\int_V (\partial F/\partial t)dV$, which would be the answer if V were fixed in space, and $\int_{\partial V} F\mathbf{u} \cdot \mathbf{n}\, dS$, which is an extra term resulting from the movement of V. Note that (2.5) is a generalization of the well-known formula for differentiating a one-dimensional integral:

$$\frac{d}{dt}\left(\int_{a(t)}^{b(t)} f(x, t)\, dx\right) = \int_{a(t)}^{b(t)} \frac{\partial f}{\partial t}\, dx + f(b, t)\frac{db}{dt} - f(a, t)\frac{da}{dt}.$$

We also remark that the function $\mathbf{u}$ in (2.5) does not have to be the velocity of the fluid everywhere inside V because we only require that $\mathbf{u} \cdot \mathbf{n}$ be the velocity of the boundary of V normal to itself.

We now apply the transport theorem to derive the equations which govern the motion of an inviscid fluid. Conservation of the mass of any material volume $V(t)$ can be written as

$$\frac{d}{dt} \left(\int_{V(t)} \rho \, dV \right) = 0,$$

where ρ is the fluid density or, using (2.4), as

$$\int_{V(t)} \left(\frac{\partial \rho}{\partial t} + \nabla \cdot (\rho \mathbf{u}) \right) dV = 0.$$

If we now shrink V to a small neighborhood of any point, we derive the differential equation

$$\frac{\partial \rho}{\partial t} + \nabla \cdot (\rho \mathbf{u}) = 0. \tag{2.6}$$

This equation is known as the *continuity equation*. We must emphasize that the above argument relies crucially on the differentiability of ρ and $\mathbf{u}$. If, as will be seen to be the case in Chapter 6, the variables are integrable but not differentiable, conservation of mass will just lead to the statement that $\int_{V(t)} \rho \, dV$ is independent of time.

We next consider the linear momentum of the fluid contained in $V(t)$. The forces created by the surrounding fluid on this volume are the "internal" surface forces exerted on the boundary ∂V, together with any "external" body forces that may be acting. If we assume that the fluid is inviscid, then the internal forces are just due to the pressure,[3] which acts along the normal to ∂V. If there is a body force $\mathbf{F}$ per unit mass and we suppose that we can apply Newton's equations to a volume of fluid, then

$$\frac{d}{dt} \left(\int_{V(t)} \rho \mathbf{u} \, dV \right) = - \int_{\partial V(t)} p \mathbf{n} \, dS + \int_{V(t)} \rho \mathbf{F} \, dV.$$

Using (2.3) on the left-hand side of this equation and the divergence theorem on the right-hand side, we obtain

$$\int_{V(t)} \left(\frac{d}{dt} (\rho \mathbf{u}) + \rho \mathbf{u} (\nabla \cdot \mathbf{u}) \right) dV = \int_{V(t)} (-\nabla p + \rho \mathbf{F}) \, dV.$$

Remembering that this is true for any volume $V(t)$ and using (2.6) leads to

$$\frac{d\mathbf{u}}{dt} = \frac{\partial \mathbf{u}}{\partial t} + (\mathbf{u} \cdot \nabla) \mathbf{u} = -\frac{1}{\rho} \nabla p + \mathbf{F}, \tag{2.7}$$

---

[3] It is at this stage that our restriction to inviscid flow is crucial. If the fluid has appreciable viscosity, the internal forces require much more careful consideration, as described in Ockendon and Ockendon [3].

which is *Euler's equation* for an inviscid fluid.[4] If (2.6) and (2.7) both hold, it can be shown that the angular momentum of any volume V is also conserved (Exercise 2.3).

For an incompressible fluid, (2.6) and (2.7) are sufficient to determine p and u, but when ρ varies, we need another relation involving p and ρ. This relation comes from considering conservation of energy, which will also involve the temperature T, thus demanding yet another relation among p, ρ, and T. When ρ is constant, the mechanical energy is automatically conserved if (2.6) and (2.7) are satisfied and there is no need to consider energy conservation unless we are concerned with thermal effects.

The energy of an inviscid compressible fluid consists of the *kinetic energy* of the fluid particles and the *internal energy* of the gas (potential energy will be accounted for separately if it is relevant). The internal energy represents the vibrational energy of the molecules of which the gas is composed and is manifested as the heat content of the gas. For an incompressible material, this heat content is the product of the specific heat and the absolute temperature, where the specific heat is determined from calorimetry. For a gas that can expand, we must take care that no unaccounted-for work is done by the pressure during the calorimetry and so we insist that the experiment is done at constant volume. The resulting specific heat is denoted by c_v.

Now, we must make a crucial assumption from thermodynamics. The *First Law of Thermodynamics* says that work, in the form of mechanical energy, can be transformed into heat, in the form of internal energy, and vice versa, without any losses being incurred. Thus, we must add the internal and mechanical energies together so that the total local "energy density" is $e + \frac{1}{2}|u|^2$, where $e = c_v T$ is the internal energy per unit mass. Now, the rate of change of energy in a material volume V must be balanced against the following:

(i) The rate at which work is done on the fluid volume by external forces.
(ii) The rate at which work is done by the body forces, and this is the term which will include the potential energy.
(iii) The rate at which heat is transferred across ∂V.
(iv) The rate at which heat is created inside V by any source terms such as radiation.

By Fourier's law, the rate at which heat is conducted in a direction n is $(-k\nabla T) \cdot n$, where k is the conductivity of the material. Thus, conservation of energy for the fluid in $V(t)$ leads to the equation

$$\frac{d}{dt}\left[\int_{V(t)} \left(\frac{1}{2}\rho|u|^2 + \rho e \right) dV \right]$$

$$= \int_V \rho F \cdot u \, dV - \int_{\partial V} p u \cdot n \, dS \int_{\partial V} k\nabla T \cdot n \, dS + \frac{d}{dt}\int_V \rho Q \, dV,$$

---

[4] Here, we use $(u \cdot \nabla)u$ to denote the operator $(u \cdot \nabla)$ in *cartesian coordinates* acting on u. In general coordinates, $(u \cdot \nabla)u$ is $\frac{1}{2}\nabla|u|^2 - u \wedge (\nabla \wedge u)$.

where Q is the heat addition per unit mass. Using the transport theorem (2.3), and (2.6) and transforming the surface integrals by the divergence theorem, we obtain the equation

$$\rho \mathbf{u} \cdot \frac{d\mathbf{u}}{dt} + \rho \frac{de}{dt} = -\nabla \cdot (p\mathbf{u}) + \rho \mathbf{F} \cdot \mathbf{u} + \nabla \cdot (k\nabla T) + \rho \frac{dQ}{dt}.$$

This can be further simplified using (2.7) and (2.6) to get

$$\rho \frac{de}{dt} = \frac{p}{\rho} \frac{d\rho}{dt} + \nabla \cdot (k\nabla T) + \rho \frac{dQ}{dt}. \qquad (2.8)$$

(see Exercise 2.2).

Looking back at (2.6), (2.7), and (2.8), we see that we have five formidable simultaneous nonlinear partial differential equations to solve. A first check shows that there are six dependent variables $\mathbf{u}$, ρ, p, and T, and, so, before we consider the appropriate boundary or initial conditions, we need to feed in some more information if we are to have any possibility of a well-posed mathematical model.

An immediate reaction is to note how much easier things are for an incompressible inviscid fluid. If we can say that ρ is constant, then the equations uncouple so that first (2.6) and (2.7) can be solved for p and $\mathbf{u}$ and (2.8) will determine T subsequently. Further than this, if we were considering a *barotropic* flow in which p is a prescribed function of ρ, then the same decomposition would occur.[5] Unfortunately, most gas flows are far from barotropic, but there is one simple relationship that holds for gases that are not being compressed or expanded too violently. This is the *perfect gas law*:

$$p = \rho R T. \qquad (2.9)$$

It is both experimentally observed and predicted from statistical mechanics arguments that R is a universal constant.[6] The law applies to gases that are not so agitated that their molecules are out of thermodynamic equilibrium. Hence if we assume that the perfect gas law does hold, we are, in effect, requiring that any non-equilibrium effects are negligible and we will discuss briefly how to model some non-equilibrium gasdynamics in Section 6.3.3 of Chapter 6. Furthermore, most observations to corroborate this law are made when the gas is at rest. This immediately raises the question of whether relation (2.9) can be used to describe the gasdynamics we are modeling here and, in particular, whether the pressure measured in static experiments can be identified with the variable p in (2.6)–(2.8). For the moment, we will simply assume that (2.9) is sufficient for practical purposes.

---

[5] Note that compressibility effects in water can be modeled by taking p proportional to ρ^γ, where γ is approximately 7; see Glass and Sislan [4].

[6] It looks strange mathematically to put this constant in between two variables, but this is the conventional notation.

We are now almost in a position to make a dramatic simplification of (2.8). Before doing so, we need one other technical result that involves two "thought experiments". Suppose first that we change the state of a constant volume V of gas from pressure p and temperature T to pressure $p + \delta p$ and temperature $T + \delta T$. We assume that the gas is in equilibrium both at the beginning and end of this experiment. Then, the amount of work needed to make this change is

$$\delta q = c_v \delta T. \tag{2.10}$$

Next, we consider changing the state by altering V and T to $V + \delta V$ and $T + \delta T$ while keeping the pressure constant. In this case, the work needed to make this change is defined to be

$$\delta q' = c_p \delta T, \tag{2.11}$$

where c_p is the specific heat at constant pressure and, from (2.9),

$$p \delta V = R \delta T. \tag{2.12}$$

Finally, we observe that if we had attained this second state from the state $p + \delta p$, $T + \delta T$, V by an isothermal (constant temperature) change, we would have had to provide an extra amount of work $p\delta V$ over and above that needed for the constant volume change. Hence,

$$\delta q' = \delta q + p\delta V$$

and so, from (2.10) and (2.11),

$$c_p \delta T = c_v \delta T + p \delta V.$$

Using (2.12), we find the relation

$$c_p - c_v = R. \tag{2.13}$$

It is conventional to define γ as the ratio of specific heats

$$\gamma = \frac{c_p}{c_v} \tag{2.14}$$

and we note that since $R > 0$, $c_p > c_v$, and so $\gamma > 1$; it can be shown from the kinetic theory of gases that $\gamma = 1.4$ for nitrogen and this is approximately the value for air under everyday conditions.

For simplicity, let us assume that there is no heat conduction by putting $k = 0$ in (2.8). (This is part of the definition of an *ideal* gas.) Then, (2.8) becomes

$$\frac{de}{dt} - \frac{p}{\rho^2}\frac{d\rho}{dt} = \frac{dQ}{dt}, \tag{2.15}$$

and we can put $e = c_v T = \frac{c_v p}{R\rho}$, on using (2.9). Now, the left-hand side of (2.15) depends only on p and ρ and we can therefore find an integrating factor that makes this expression proportional to a total derivative. A simple calculation using (2.13) and (2.14) shows that

$$\frac{de}{dt} - \frac{p}{\rho^2}\frac{d\rho}{dt} = \frac{c_v}{R\rho}\frac{dp}{dt} - \left(\frac{c_v p}{R\rho^2} + \frac{p}{\rho^2}\right)\frac{d\rho}{dt}$$

$$= \frac{c_v}{R\rho}\left[\frac{dp}{dt} - \frac{\gamma p}{\rho}\frac{d\rho}{dt}\right]$$

$$= c_v T\frac{d}{dt}\left(\log\frac{p}{\rho^\gamma}\right).$$

Hence, if we write $S = S_0 + c_v \log(p/\rho^\gamma)$, where S_0 is a constant, we obtain the celebrated result

$$T\frac{dS}{dt} = \frac{dQ}{dt}. \tag{2.16}$$

The formal relation $T\delta S = \delta Q$ is the usual starting point for the definition of the *entropy* S of a gas; when a unit mass of gas is heated by an amount δQ, its entropy is defined to be a function that changes by $\delta Q/T$. However, by starting from the energy equation, we have shown that this mysterious function arises quite naturally in gasdynamics. The above discussion also enables us to state at once that since volumetric radiative cooling with $\delta Q < 0$ has never been observed experimentally, and since $T \geq 0$, then $dS/dt \geq 0$, which is a manifestation of the *Second Law of Thermodynamics*.

Finally, reinstating the conduction term in the energy equation, we can write (2.8) as

$$T\frac{dS}{dt} = \frac{1}{\rho}\nabla \cdot (k\nabla T) + \frac{dQ}{dt}. \tag{2.17}$$

In most of the subsequent work, k and Q will be taken to be zero and so the equation will reduce to

$$\frac{dS}{dt} = 0. \tag{2.18}$$

In this situation, S is constant for a fluid particle and the flow is *isentropic*. If, in addition, the entropy of *all* fluid particles is the same (as would happen if the gas was initially uniform for instance), then $S \equiv S_0$ and the flow is *homentropic*.

In fact, the Second Law of Thermodynamics states that the *total* entropy of any thermodynamical system can never decrease, but here we have obtained the stronger statement (2.18) that the rate of change of entropy of any fluid particle is zero. Now, it is well known (see, e.g., Ockendon and Ockendon [3], that in any viscous flow in which there is shear, there is a positive dissipation of mechanical to thermal energy. Hence, we expect dS/dt to be positive whenever viscosity is present. On the other hand, as shown in Exercise 2.6, thermal conduction is a less powerful dissipative mechanism than viscosity because the equation $T(dS/dt) = (1/\rho)\nabla \cdot (k\nabla T)$ does not constrain the sign of dS/dt.[7] We will return to these ideas in more detail in Chapter 6.

We have now succeeded in writing down six equations [(2.6), (2.7), (2.9), and (2.16)], for our six dependent variables. Before considering their implications, we will consider briefly the sort of initial and boundary conditions that may arise.

2.2 Initial and Boundary Conditions

The presence of a single time derivative in each of (2.6)–(2.8) suggests that no matter what the boundary conditions are, we will require initial values for ρ, $\mathbf{u}$, and T and these will give the initial value for p from (2.9).

The boundary conditions are easy enough to guess when there is a prescribed impermeable boundary to the flow. We simply synthesize what is known about incompressible inviscid flow and what is known about heat conduction in solids to propose the following:

(i) The kinematic condition: The normal component of $\mathbf{u}$ should be equal to the normal velocity of the boundary (with no condition on p).
(ii) The thermodynamic condition: The temperature or the heat flux, $-k\mathbf{n} \cdot \nabla T$, or some combination of these two quantities should be prescribed. This assumes that $k > 0$; if $k = 0$, then no thermodynamic condition is needed.

For a prescribed, moving, impermeable boundary $f(\mathbf{x}, t) = 0$, we note that a consequence of the assumption that the gas is a continuum is that fluid particles which are on the boundary of a fluid at any time must always remain on the boundary. Hence, the kinematic condition on the boundary is

$$\frac{df}{dt} = 0 = \frac{\partial f}{\partial t} + \mathbf{u} \cdot \nabla f. \tag{2.19}$$

However, the situation becomes much more complicated when the boundary of the gas is free rather than being prescribed. This could occur if the gas was confined behind a shock wave and this difficult situation will be discussed in Chapter 6. Things are simpler for an incompressible flow, such as the flow of water with a free surface; now, we must impose a second condition over and above the kinematic condition (2.19) if we are to be able to solve the field equations and *also* determine the position of the boundary. This second condition comes from considering the momentum balance. A simple argument suggests

---

[7] We hasten to emphasize that in most gases, the effects of viscosity and thermal conductivity are of comparable size. Hence, the study of an inviscid gas with $k > 0$ is of purely academic interest.

that in the absence of surface tension, the pressure must be continuous across the boundary, because the boundary has no inertia; hence,

$$p_1 = p_2 \qquad (2.20)$$

on the boundary, where p_2 is the external pressure and p_1 is the pressure in the fluid. Conditions (2.19) and (2.20) will be reconsidered more carefully in specific circumstances in later chapters.

Before considering the full implications of the model we have derived, it is very helpful to recall some well-known results about vorticity, circulation and incompressible flow. This will not only help us pose the best questions to ask about compressible flows in general but will also provide useful background for some of the models to be considered in Chapter 3.

2.3 Vorticity and Irrotationality

2.3.1 Homentropic Flow

One distinctive attribute of fluid mechanics, compressible or incompressible, compared to other branches of continuum mechanics is the existence of *vorticity ω*, defined by $\omega = \nabla \wedge \mathbf{u}$. We can derive an equation for the evolution of ω by first writing

$$(\mathbf{u} \cdot \nabla)\mathbf{u} = \tfrac{1}{2}\nabla|\mathbf{u}|^2 - \mathbf{u} \wedge (\nabla \wedge \mathbf{u})$$

in (2.7). If we assume that $\mathbf{F}$ is a conservative force so that $\mathbf{F} = -\nabla\Omega$ for some scalar potential Ω and we use the same algebraic manipulations as those used to derive (2.16), we obtain

$$\frac{d\mathbf{u}}{dt} = \frac{\partial \mathbf{u}}{\partial t} + \nabla\left(\frac{1}{2}|\mathbf{u}|^2\right) - \mathbf{u} \wedge \omega = \nabla\left(-\Omega - \frac{\gamma p}{(\gamma-1)\rho}\right) + T\nabla S. \qquad (2.21)$$

Taking the curl of this equation leads to

$$\frac{\partial \omega}{\partial t} + (\mathbf{u} \cdot \nabla)\omega = (\omega \cdot \nabla)\mathbf{u} + \nabla \wedge (T\nabla S),$$

or

$$\frac{d\omega}{dt} = (\omega \cdot \nabla)\mathbf{u} + \nabla T \wedge \nabla S.$$

For a homentropic fluid, ∇S will be zero and so the equation for ω is then

$$\frac{d\omega}{dt} = (\omega \cdot \nabla)\mathbf{u}. \qquad (2.22)$$

Thus, in two-dimensional homentropic flow, in which $(\omega \cdot \nabla)\mathbf{u}$ is automatically zero, vorticity is convected with the fluid. Remarkably, if we change to

Lagrangian variables, (2.22) can be solved explicitly, even in three dimensions (see Exercise 2.4), to give

$$\boldsymbol{\omega} = (\boldsymbol{\omega}_0 \cdot \nabla_{\mathbf{a}})\mathbf{x}, \tag{2.23}$$

where $\nabla_{\mathbf{a}}$ is the gradient operator with respect to Lagrangian variables $\mathbf{a}$, and $\boldsymbol{\omega}_0$ is the value of $\boldsymbol{\omega}$ at $t = 0$. This is *Cauchy's equation* for the vorticity in an arbitrary homentropic flow, but it is not very useful since we cannot find $\nabla_{\mathbf{a}}$ until we have found the flow field! However, (2.23) does tell us immediately that if the vorticity is everywhere zero in a fluid region $V(0)$ at $t = 0$, then it will be zero at all subsequent times in the region $V(t)$, which contains the same fluid particles as $V(0)$. Thus, $\boldsymbol{\omega} \equiv \mathbf{0}$ in $V(t)$ and the flow is *irrotational*. Such flows occur, for example, when the fluid is initially at rest or when there are uniform conditions at infinity in steady flow.

To understand vorticity transport geometrically, we plot the trajectories of two nearby fluid particles that are at $\mathbf{x}(t)$ and $\mathbf{x}(t) + \varepsilon\boldsymbol{\omega}(\mathbf{x}(t), t)$ at time t, as shown in Figure 2.1. After a short time δt, the particles will have moved to $\mathbf{x}(t) + \mathbf{u}(\mathbf{x}, t)\delta t$ and $\mathbf{x}(t) + \varepsilon\boldsymbol{\omega}(\mathbf{x}, t) + \mathbf{u}(\mathbf{x} + \varepsilon\boldsymbol{\omega}(\mathbf{x}, t), t)\delta t$, respectively, and the vector joining the two particles will therefore have changed from $\varepsilon\boldsymbol{\omega}(\mathbf{x}, t)$ to $\varepsilon\boldsymbol{\omega}(\mathbf{x}, t) + \varepsilon(\boldsymbol{\omega} \cdot \nabla)\mathbf{u}\delta t$. However, from (2.22),

$$(\boldsymbol{\omega} \cdot \nabla)\mathbf{u}\delta t = \boldsymbol{\omega}(\mathbf{x}(t + \delta t), t + \delta t) - \boldsymbol{\omega}(\mathbf{x}(t), t),$$

and so the vector joining the particles at $t + \delta t$ is $\varepsilon\boldsymbol{\omega}(\mathbf{x}(t + \delta t), t + \delta t)$. Thus, we can see that, in three dimensions, the *vortex lines*, which are parallel to the vorticity at each point of the fluid, move with the fluid and are stretched as the vorticity increases.

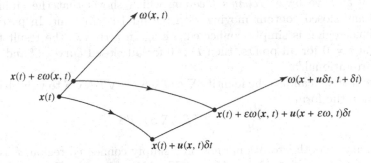

Fig. 2.1. Convection of vorticity.

An alternative way to approach vorticity is to consider the total vorticity flux through an arbitrary closed contour $C(t)$ which moves with the fluid. This quantity, known as the *circulation* around C, is given by

$$\Gamma = \int_C \mathbf{u} \cdot d\mathbf{x} = \int_\Sigma \boldsymbol{\omega} \cdot d\mathbf{S},$$

where Σ is any smooth surface spanning C and contained within the fluid. Note that the circulation integral around C is defined even in a non-simply connected region. To consider the rate of change of Γ, we change to Lagrangian variables so that

$$\Gamma = \int_{C(t)} u_i \, dx_i = \int_{C(0)} u_i \frac{\partial x_i}{\partial a_j} \, da_j.$$

Then,

$$\frac{d\Gamma}{dt} = \int_{C(0)} \frac{du_i}{dt} \frac{\partial x_i}{\partial a_j} \, da_j + u_i \frac{\partial \dot{x}_i}{\partial a_j} \, da_j$$

$$= \int_{C(t)} \left(\frac{d\mathbf{u}}{dt} \cdot d\mathbf{x} + \mathbf{u} \cdot d\mathbf{u} \right).$$

Now,

$$\int_C \mathbf{u} \cdot d\mathbf{u} = [\tfrac{1}{2}(\mathbf{u})^2]_C = 0$$

since $\mathbf{u}$ is a single-valued function, and, so, using (2.21),

$$\frac{d\Gamma}{dt} = \int_{C(t)} T\nabla S \cdot d\mathbf{x} - \left[\Omega + \frac{\gamma p}{(\gamma - 1)\rho} \right]_{C(t)}$$

$$= \int_{C(t)} T\nabla S \cdot d\mathbf{x},$$

since Ω, p, and ρ are all single-valued functions. For a homentropic flow, $\nabla S = \mathbf{0}$ and we have *Kelvin's theorem*, which shows that the circulation around any closed contour moving with the fluid is constant. In particular, if the fluid region is simply connected, we again arrive at the result that if $\boldsymbol{\omega} \equiv \mathbf{0}$ at $t = 0$ for all points, then $\Gamma \equiv 0$ for all closed curves C and so the flow is irrotational.[8]

Note that we can use the identity $\nabla \wedge (T\nabla S) = \nabla T \wedge \nabla S$ to write Kelvin's theorem in the form

$$\frac{d\Gamma}{dt} = \int_\Sigma (\nabla T \wedge \nabla S) \cdot d\mathbf{S}.$$

Now, in any smooth irrotational flow in a simply connected region, Γ is identically zero and so, since Σ is arbitrary, $\nabla T \wedge \nabla S = \mathbf{0}$. Since T is proportional to p/ρ and S is a function of p/ρ^γ with $\gamma > 1$, T cannot be a function of S alone and so the flow must be either homentropic or isothermal. The latter is unlikely in practice, and vorticity can thus be associated with an entropy gradient and vice versa except in special cases (see Exercise 2.5).

---

[8] It is easy to see that this result does not apply in, say, a circular annulus when $\mathbf{u} = (\Gamma/2\pi)\mathbf{e}_\theta$ in polar coordinates.

Whenever the flow is irrotational, we can define a *velocity potential* ϕ by $\phi(\mathbf{x}, t) = \int_{\mathbf{x}_0}^{\mathbf{x}} \mathbf{u} \cdot d\mathbf{x}$ for any convenient constant $\mathbf{x}_0$, and from Kelvin's theorem, ϕ will be a well-defined function of $\mathbf{x}$ and t. From this definition, we can write

$$\mathbf{u} = \nabla \phi.$$

Now, substituting for $\mathbf{u}$ in (2.6) and (2.21), the equations for homentropic irrotational flow with a conservative body force collapse to

$$\frac{\partial \rho}{\partial t} + \nabla \cdot (\rho \nabla \phi) = 0 \qquad (2.24)$$

and

$$\frac{\partial \phi}{\partial t} + \frac{1}{2}|\nabla \phi|^2 + \Omega + \frac{\gamma p}{(\gamma - 1)\rho} = G(t), \qquad (2.25)$$

where G is some function of t, often determined by the conditions at infinity. Equation (2.25) is *Bernoulli's equation* for homentropic gas flow.

2.3.2 Incompressible Flow

Most of the modeling in the previous section is an obvious generalization of well-known results for inviscid incompressible flows. In particular, homentropic compressible flow has many features in common with incompressible flow; (2.22) and (2.23) hold for incompressible flow, as does Kelvin's theorem, and in both cases, the existence of a velocity potential in irrotational flow leads to a dramatic simplification.

However, the incompressible limit of our compressible model is non-trivial mathematically and we only make one general remark about it here, although we will return to it again in Chapter 4. In the light of footnote 5 on page 10, one possible procedure is to let $\gamma \to \infty$. Now, γ only enters the general model via the energy equation in the form (2.18), which we can write as

$$\frac{d}{dt}\left(\frac{\rho}{p^{1/\gamma}}\right) = 0.$$

Letting $\gamma \to \infty$ now clearly suggests that $d\rho/dt = 0$ and, hence, that the flow is incompressible. We also note that letting $\gamma \to \infty$ in (2.25) leads to the familiar incompressible form of Bernoulli's equation.

We will now use our *nonlinear* model for gasdynamics as a basis for the *linearized* theory of *acoustics* or *sound waves*. This will lead us to the prototype of all models for wave motion. Even more importantly, it will show how the linearization of an intractable nonlinear problem can lead to a linear wave propagation model which is both revealing and straightforward to analyze.

Exercises

R2.1 If J is the Jacobian $\partial(x_1, x_2, x_3)/\partial(a_1, a_2, a_3)$, where $\mathbf{a}$ are Lagrangian coordinates, use (2.2) and (2.6) to show that $d(\rho J)/dt = 0$.

R2.2 The equations for a compressible gas are, in the absence of heat conduction or radiation,

$$\frac{\partial \rho}{\partial t} + \nabla \cdot (\rho \mathbf{u}) = 0, \tag{1}$$

$$\frac{\partial}{\partial t}(\rho \mathbf{u}) + (\mathbf{u} \cdot \nabla)(\rho \mathbf{u}) + \rho \mathbf{u}(\nabla \cdot \mathbf{u}) = -\nabla p \tag{2}$$

and

$$\frac{\partial}{\partial t}\left(\rho \left(e + \frac{1}{2}|\mathbf{u}|^2\right)\right) + \nabla \cdot \left(\rho \left(e + \frac{1}{2}|\mathbf{u}|^2\right)\mathbf{u}\right) = -\nabla \cdot (p\mathbf{u}). \tag{3}$$

From (1) and (2) show that the Euler equation

$$\frac{d\mathbf{u}}{dt} = -\frac{1}{\rho}\nabla p \tag{4}$$

holds. Using (1) and (4) to eliminate $d\rho/dt$ and $d\mathbf{u}/dt$ from (3), show that

$$\rho \frac{de}{dt} = -p\nabla.\mathbf{u},$$

and, hence, from (1) that

$$\frac{de}{dt} = \frac{p}{\rho^2}\frac{d\rho}{dt}.$$

Deduce that p/ρ^γ is a constant for a fluid particle in a perfect gas.

2.3 Define the angular momentum of a material volume V as

$$\mathbf{L} = \int_{V(t)} \mathbf{x} \wedge \rho \mathbf{u}\, dV,$$

where $\mathbf{x}$ is the position of a particle of fluid with respect to a fixed origin. Show by using (2.6) and (2.7) that

$$\frac{d\mathbf{L}}{dt} = -\int_{\partial V(t)} \mathbf{x} \wedge p\mathbf{n}\, dS + \int_{V(t)} \mathbf{x} \wedge \rho \mathbf{F}\, dV$$

and deduce that the rate of change of angular momentum of the fluid in $V(t)$ is equal to the sum of the moments of the forces acting on $V(t)$.

Note that if this formula is applied to the angular momentum of a small element of fluid Σ about its center of gravity, the magnitude of $\mathbf{L}$ will be of $O(\delta^4)$ if δ is the length scale of the element, whereas the term $\int \mathbf{x} \wedge p\mathbf{n}\, dS$ is of $O(\delta^3)$. Formally, letting $\delta \to 0$ gives

$$p\int_{\partial \Sigma} \mathbf{x} \wedge \mathbf{n}\, dS = 0,$$

which, fortunately, is identically true.

2.4 Starting from the Euler equation (2.7) with $\mathbf{F} = \mathbf{0}$, show that, in homentropic flow, the vorticity $\boldsymbol{\omega} = \nabla \wedge \mathbf{u}$ satisfies the equation

$$\frac{d\boldsymbol{\omega}}{dt} = (\boldsymbol{\omega} \cdot \nabla)\mathbf{u}.$$

By changing to Lagrangian variables $\mathbf{a}$, and t, where $\mathbf{x}(\mathbf{a}, 0) = \mathbf{a}$, show that

$$\frac{d\omega_i}{dt} = \omega_k \frac{\partial a_j}{\partial x_k} \frac{d}{dt}\left(\frac{\partial x_i}{\partial a_j}\right),$$

where the summation convention for the repeated suffices j and k is used. Noting that $(\partial x_i/\partial a_k) \cdot (\partial a_k/\partial x_j) = \delta_{ij}$, show that

$$\frac{d}{dt}\left(\omega_k \frac{\partial a_i}{\partial x_k}\right) = 0$$

and, hence, deduce that

$$\boldsymbol{\omega} = (\boldsymbol{\omega}_0 \cdot \nabla_{\mathbf{a}})\mathbf{x},$$

where $\boldsymbol{\omega} = \boldsymbol{\omega}_0$ at $t = 0$.

2.5 Show that in a *two-dimensional* steady flow, the entropy S is constant on a streamline and, hence that $\mathbf{u}$ and $\nabla \wedge \mathbf{u}$ are perpendicular to ∇S. Deduce *Crocco's theorem*, which states that for rotational, non-homentropic flow,

$$\mathbf{u} \wedge (\nabla \wedge \mathbf{u}) = \lambda \nabla S$$

for some scalar function λ.

Show that for the steady two-dimensional flow $\mathbf{u} = (y, 0, 0)$, the entropy S must be a function of y and, hence that it is possible for a rotational flow to be homentropic. Show also that for the three-dimensional rotational flow $\mathbf{u} = (0, \cos x, -\sin x)$, it is again possible for the flow to be homentropic.

2.6 Show that in a heat conducting gas with positive conductivity k (which need not be constant),

$$T\frac{dS}{dt} = \frac{1}{\rho}\nabla \cdot (k\nabla T).$$

Deduce that if the gas is confined in a fixed thermally insulated container Ω, then the rate of change of total entropy is

$$\frac{d}{dt}\left[\int_\Omega \rho S\, dV\right] = \int_\Omega \frac{k|\nabla T|^2}{T^2}\, dV \geq 0.$$

2.7 If Ω is an arbitrary volume of fluid *fixed in space*, show that the principle of conservation of mass implies that

$$\frac{d}{dt}\int_\Omega \rho\, dV = -\int_{\partial\Omega} \rho\mathbf{u} \cdot d\mathbf{S}$$

and hence deduce (2.6). In a similar way, deduce (2.7) and (2.8) by considering the momentum and energy of the fluid in Ω.

3

Models for Linear Wave Propagation

This chapter will discuss models for several quite different classes of waves with the common characteristic that they are of sufficiently small amplitude for the models to be linear. We will focus on waves in fluids, but even here, we will find that the models are far from trivial and can look very different from each other. Their unifying features will become more apparent when we embark on their mathematical analysis in Chapter 4. We begin with sound waves, which are one of the most familiar of all waves.

3.1 Acoustics

The theory of acoustics is based on the fact that in sound waves (at least those that do not affect the eardrum adversely), the variations in pressure, density, and temperature are all small compared to some ambient conditions. These ambient conditions from which the motion is initiated are usually either that the gas is at rest, so that $p = p_0$, $\rho = \rho_0$, $T = T_0$, and $\mathbf{u} = \mathbf{0}$, or the gas is in a state of uniform motion in which $\mathbf{u} = U\mathbf{i}$, say. We start with the simplest case and motivate the linearization procedure in an intuitive way.

We suppose that the gas is initially at rest in a long pipe along the x axis and that it is subject to a small disturbance so that

$$\mathbf{u} = \bar{u}(x, t)\mathbf{i}.$$

We assume that $\bar{p} = p - p_0$ and $\bar{\rho} = \rho - \rho_0$ are "small" and neglect the squares of the barred quantities. From (2.6) and (2.7), we find

$$\frac{\partial \bar{\rho}}{\partial t} + \rho_0 \frac{\partial \bar{u}}{\partial x} = 0 \tag{3.1}$$

and

$$\frac{\partial \bar{u}}{\partial t} + \frac{1}{\rho_0} \frac{\partial \bar{p}}{\partial x} = 0. \tag{3.2}$$

The energy equation (2.17) reduces to $p/\rho^\gamma = p_0/\rho_0^\gamma$, so that

$$\bar{p} = \frac{\gamma p_0}{\rho_0}\bar{\rho} \qquad (3.3)$$

to a first approximation. We define c_0^2 to be $\gamma p_0/\rho_0$, and then, from (3.1), (3.2), and (3.3), we can show that the variables $\bar{\rho}$, $\bar{u}$, and $\bar{p}$ all satisfy the same equation, namely

$$\frac{\partial^2 \phi}{\partial x^2} = \frac{1}{c_0^2}\frac{\partial^2 \phi}{\partial t^2}; \qquad (3.4)$$

this is the well-known one-dimensional wave equation which generates waves traveling with speed $\pm c_0$, and c_0 is known as the *speed of sound*.

The simplicity of (3.4) in comparison with (2.6)–(2.8) is dramatic and the validity of the linearization procedure requires careful scrutiny. In fact, even assuming that we are in a regime where (2.6), (2.7), and (2.8) are valid, much more care is needed to derive (3.4) than the simple assumption that the square of the perturbations (the barred variables) can be neglected. Most strikingly, even though $\bar{u}$ is small, $\partial \bar{u}/\partial x$ may be large, so that the neglect of the nonlinear term $\bar{u}(\partial \bar{u}/\partial x)$ may not be justified. Also, not only must the amplitude of the waves be small, but the time variation must not be too slow if it is to interact with the spatial variation. In order to clarify the assumptions built into the approximation represented by (3.4), we need to do a systematic non-dimensionalization and analyze the equations as below.

In many circumstances, the wave motion will be driven with a prescribed velocity u_0, and frequency ω_0 and propagate over a known length scale L. We therefore introduce non-dimensional variables

$$\rho = \rho_0(1 + \varepsilon\hat{\rho}),$$
$$p = p_0(1 + \varepsilon\hat{p}),$$
$$u = u_0\hat{u},$$
$$x = LX$$

and

$$t = \omega_0^{-1}T,$$

where ε is a small dimensionless parameter. Then, (2.6) and (2.7) become

$$\varepsilon\rho_0\omega_0\frac{\partial\hat{\rho}}{\partial T} + \frac{\rho_0 u_0}{L}\frac{\partial\hat{u}}{\partial X} + \frac{\rho_0 u_0 \varepsilon}{L}\frac{\partial}{\partial X}(\hat{u}\hat{\rho}) = 0$$

and

$$(1 + \varepsilon\hat{\rho})\left(u_0\omega_0\frac{\partial\hat{u}}{\partial T} + \frac{u_0^2}{L}\hat{u}\frac{\partial\hat{u}}{\partial X}\right) = -\frac{\varepsilon p_0}{L\rho_0}\frac{\partial\hat{p}}{\partial X}.$$

These equations will thus retain the same terms as (3.1) and (3.2), as a first approximation in ε, if[1] $u_0 \simeq \varepsilon\omega_0 L \simeq \varepsilon p_0/L\omega_0\rho_0$ and, remembering that $c_0^2 =$

---

[1] Here, we use the symbol $\simeq$ to mean "is approximately equal to."

$\gamma p_0/\rho_0$, this is achieved by taking

$$\omega_0 L \simeq c_0, \quad u_0 \simeq \varepsilon c_0. \tag{3.5}$$

Thus if, for example, the motion is being driven by a piston oscillating with speed u_0, then u_0 must be much smaller than the speed of sound in the undisturbed gas for the linearization to be valid. If ε is defined to be u_0/c_0, then the resulting pressure and density variations will automatically be of $O(\varepsilon p_0)$ and $O(\varepsilon \rho_0)$. Equally, if the motion is driven by a prescribed pressure oscillation of amplitude $O(\varepsilon p_0)$, then the resulting density and velocity changes will be $O(\varepsilon \rho_0)$ and $O(\varepsilon c_0)$. In all cases, our theory will only describe waves whose frequency is no higher than $O(c_0/L)$.

Although this derivation of (3.4) is more laborious than the simple hand-waving that we used at the beginning of the section, it is the only way we can have any reliable knowledge of the range of validity of the model and we will need to take this degree of care throughout this chapter.

We note here some other important but less fundamental remarks about the acoustic approximation.

(i) **Sound waves in three dimensions.** As shown in Exercise 3.1, in higher dimensions, (3.4) is replaced by[2]

$$\nabla^2 \phi = \frac{1}{c_0^2} \frac{\partial^2 \phi}{\partial t^2}. \tag{3.6}$$

This may still reduce to a problem in two variables if we have either circular symmetry, when $\nabla^2 = \partial^2/\partial r^2 + (1/r)(\partial/\partial r)$, or spherical symmetry, when $\nabla^2 = \frac{\partial^2}{\partial r^2} + \frac{2}{r}\frac{\partial}{\partial r}$, in suitable polar coordinates.

(ii) **Sound waves in a medium moving with uniform speed U.** If the uniform flow U is taken along the x axis, it can be shown (Exercise 3.1) that by writing $\mathbf{u} = U\mathbf{i} + \varepsilon\nabla\phi$, (3.4) is now replaced by

$$\nabla^2 \phi = \frac{1}{c_0^2} \left(\frac{\partial}{\partial t} + U\frac{\partial}{\partial x} \right)^2 \phi.$$

In particular, for steady flow,

$$\left(1 - \frac{U^2}{c_0^2} \right) \frac{\partial^2 \phi}{\partial x^2} + \frac{\partial^2 \phi}{\partial y^2} + \frac{\partial^2 \phi}{\partial z^2} = 0, \tag{3.7}$$

and it is clear that the parameter U/c_0 now plays a key role in the solution. It is called the *Mach number* and the flow is *supersonic* if $M > 1$ ($U > c_0$) and *subsonic* if $M < 1$ ($U < c_0$). Note that the Mach number of acoustic waves in a stationary medium is of $O(\varepsilon)$ by (3.5), even though the waves themselves propagate at sonic speed.

---

[2] Note that the three-dimensional version of (3.2) is $\partial\mathbf{u}/\partial t = -(1/\rho_0)\nabla p$, which automatically guarantees that $\partial\omega/\partial t = 0$; this makes irrotationality even more common than Kelvin's theorem suggests.

3.2 Surface Gravity Waves in Incompressible Flow

We now consider the problem of waves on the surface of an incompressible fluid subject to gravitational forces. It may seem strange to suddenly revert to incompressible flow at this stage, but, in fact, we can think of water and air separated by an interface as an extreme case of a variable density fluid where all the density variation takes place at the surface. The ratio of densities of air and water is about 10^{-3}, so the jump is extreme in magnitude as well as occurring over a very short distance. We will come back to this point of view later, but for the moment, we will derive the governing equations from the usual equations of incompressible fluid dynamics.

We recalled in Chapter 2 that the classical theory of inviscid flow predicts that if the fluid motion is initially irrotational, then it will remain irrotational. Thus, writing $\mathbf{u} = \nabla\phi$, the field equations reduce to Laplace's equation

$$\nabla^2\phi = 0 \tag{3.8}$$

for ϕ and to Bernoulli's equation

$$\frac{\partial\phi}{\partial t} + \frac{1}{2}|\nabla\phi|^2 + gz + \frac{p}{\rho} = \frac{p_0}{\rho} \tag{3.9}$$

for p, where we have assumed that the external pressure in the air is p_0 and that the z axis is vertical. What is important now are the boundary conditions for ϕ at the free surface. We anticipate that whereas only one condition is needed for ϕ at a prescribed boundary, we will now need two conditions to compensate for the fact that the position of the free surface is unknown and needs to be determined as part of the solution of the problem. A problem of this type is known as a *free boundary problem*.

The first free surface condition comes from the fact that no fluid particle can cross the surface (we will neglect any "spray"). If the surface is given by $z = \eta(x, t)$, where we are considering a two-dimensional situation for simplicity, a particle on the surface has position $(x, 0, \eta)$ and the velocity of this particle is $(u, 0, w)$, where

$$w = \frac{d\eta}{dt} = \frac{\partial\eta}{\partial t} + u\frac{\partial\eta}{\partial x}.$$

Hence, as could have also been deduced from (2.19), we have the *kinematic* boundary condition

$$\frac{\partial\phi}{\partial z} = \frac{\partial\eta}{\partial t} + \frac{\partial\phi}{\partial x}\cdot\frac{\partial\eta}{\partial x}, \tag{3.10}$$

which expresses the principle of conservation of mass at the free surface.

The second condition expresses the principle of conservation of momentum at the free surface. As discussed in Chapter 2, this simply means that, if surface tension effects can be neglected,[3] then the pressure at the surface will be p_0, so that, from (3.9),

$$\frac{\partial \phi}{\partial t} + \frac{1}{2}|\nabla \phi|^2 + g\eta = 0 \tag{3.11}$$

on $z = \eta(x, t)$.

If we apply suitable initial conditions (which must satisfy irrotationality) and conditions at any fixed boundaries, we will have a fully nonlinear model for surface gravity waves. This model is every bit as formidable as the compressible equations (2.6), (2.7), and (2.18), so let us again consider the effect of linearization. We will take water of depth h at rest as the basic equilibrium state and formally neglect squares and products of the variables ϕ and η. There is one extra subtlety here because when we make this assumption in (3.10), we must, to be consistent, write

$$\frac{\partial \phi}{\partial z} = \frac{\partial \eta}{\partial t} \quad \text{on } z = 0,$$

rather than on $z = \eta$. This is because the difference between $\partial \phi(x, \eta, t)/\partial z$ and $\partial \phi(x, 0, t)/\partial z$ is a product of η and $\partial^2 \phi/\partial z^2$ and thus is negligible under the linearization approximation. Hence, from (3.8), (3.10), and (3.11), the formal model for small-amplitude waves, called *Stokes waves*, on water of depth h is

$$\nabla^2 \phi = 0, \tag{3.12}$$

with

$$\frac{\partial \phi}{\partial z} = \frac{\partial \eta}{\partial t}, \quad \frac{\partial \phi}{\partial t} + g\eta = 0 \quad \text{on } z = 0 \tag{3.13}$$

and

$$\frac{\partial \phi}{\partial z} = 0 \quad \text{on } z = -h. \tag{3.14}$$

The conditions (3.13) can be further reduced to a single condition on ϕ in the form

$$\frac{\partial^2 \phi}{\partial t^2} + g\frac{\partial \phi}{\partial z} = 0 \quad \text{on } z = 0 \tag{3.15}$$

and we are left with the problem of solving Laplaces equation (3.8) with an odd-looking boundary condition (3.15) on one prescribed boundary and a more standard condition (3.14) on the other. Although linearization has greatly simplified the difficulty caused by the free boundary, (3.15) poses a new challenge. Standard theory tells us that Laplace's equation can usually be solved uniquely, or to within a constant, if ϕ or its normal derivative or even a linear combination thereof is prescribed on the boundary of a closed region, but (3.15) does not fall into any of these categories.

Before making any further remarks about this model, we will repeat the procedure adopted in Section 3.1 for discussing the parameter regime in which

---

[3] See Exercise 4.5 of Chapter 4 for a brief discussion of the effect of surface tension.

we might expect (3.12)–(3.15) to be valid. We suppose that the disturbance to the surface of the water has an amplitude a, which must be small compared to the depth h. Then, we non-dimensionalize by introducing an arbitrary length scale λ, time scale ω_0^{-1}, and potential scale ϕ_0 and writing $\eta = a\hat{\eta}$, $x = \lambda X$, $z = \lambda Z$, $t = \omega_0^{-1} T$, and $\phi = \phi_0 \hat{\phi}$. We find that the linearized equations (3.12)–(3.15) are a valid approximation provided λ, ω_0, and ϕ_0 satisfy

$$\omega_0 = \left(\frac{g}{\lambda}\right)^{1/2}, \quad \phi_0 = a(\lambda g)^{1/2}, \quad \text{and} \quad \frac{a}{\lambda} \ll 1. \tag{3.16}$$

Since the boundary condition (3.14) is applied on $Z = -h/\lambda$, we will also need to insist that $h/\lambda \geq O(1)$. If this latter restriction is violated, we can still make simplifications, and these lead to the nonlinear *shallow water* theory, as will be described in Chapter 5.

Once again, we can extend this theory easily enough to three dimensions when (3.12)–(3.15) will still be valid as long as we write $\nabla^2 \phi$ as $\partial^2 \phi/\partial x^2 + \partial^2 \phi/\partial y^2 + \partial^2 \phi/\partial z^2$. It is also straightforward to consider waves on a uniform stream moving with velocity U_i and in this case, the only change is that (3.15) becomes

$$\left(\frac{\partial}{\partial t} + U\frac{\partial}{\partial x}\right)^2 \phi + g\frac{\partial \phi}{\partial z} = 0.$$

3.3 Inertial Waves

As a generalization of the last section, we now consider flows which consist of incompressible particles but where the density may vary from particle to particle. This may arise, for example, in oceanography, where the density of the sea is related to the salinity, and diffusion is so small that the salinity of a fluid particle is conserved. Thus,

$$\frac{d\rho}{dt} = \frac{\partial \rho}{\partial t} + \mathbf{u} \cdot \nabla \rho = 0, \tag{3.17}$$

and (2.6) and (2.7) reduce to

$$\nabla \cdot \mathbf{u} = 0 \tag{3.18}$$

and

$$\frac{\partial \mathbf{u}}{\partial t} + (\mathbf{u} \cdot \nabla)\mathbf{u} = -\frac{1}{\rho}\nabla p - g\mathbf{k}, \tag{3.19}$$

where $\mathbf{k}$ is measured vertically upward. We now have sufficient equations to solve for $\mathbf{u}$, p, and ρ. Moreover, using (2.7) in the energy equation removes the terms involving the gravitational body force and reduces (2.8) to

$$\frac{de}{dt} = 0.$$

Thus, when there is no conduction, the temperature is constant for each fluid particle.

An exact hydrostatic solution of (3.17)–(3.19) is that of a *stratified* fluid where

$$\mathbf{u} = 0, \qquad \rho = \rho_s(z), \quad \text{and} \quad p = p_0 - g\int_0^z \rho_s(\sigma)\,d\sigma = p_s(z), \qquad (3.20)$$

say, where p_0 is a constant reference pressure on $z = 0$. Now, we can, as usual, effect a handwaving derivation of the linear theory about the state given by (3.20). For simplicity, we look at two-dimensional disturbances and assume that $\bar{\rho} = \rho - \rho_s(z)$, $\bar{p} = p - p_s(z)$, and $|\mathbf{u}| = |(u,0,w)|$ are all small. Then, with $\frac{d}{dz}$ denoted by a prime, (3.17)–(3.19) reduce to

$$\frac{\partial\bar{\rho}}{\partial t} + \rho_s' w = 0, \qquad (3.21)$$

$$\frac{\partial u}{\partial x} + \frac{\partial w}{\partial z} = 0, \qquad (3.22)$$

$$\rho_s\frac{\partial u}{\partial t} = -\frac{\partial\bar{p}}{\partial x} \qquad (3.23)$$

and

$$\rho_s\frac{\partial w}{\partial t} = -\frac{\partial\bar{p}}{\partial z} - \bar{\rho}g. \qquad (3.24)$$

It is now a simple matter to cross-differentiate to eliminate $\bar{\rho}$, $\bar{p}$, and u to obtain

$$\frac{\partial^2}{\partial t^2}\left(\frac{\partial^2 w}{\partial x^2} + \frac{\partial^2 w}{\partial z^2}\right) = -N^2(z)\left(\frac{\partial^2 w}{\partial x^2} - g^{-1}\frac{\partial^3 w}{\partial z\partial t^2}\right), \qquad (3.25)$$

where $N^2(z) = -g\rho_s'(z)/\rho_s(z)$ is a positive function in a stably stratified fluid. We note with satisfaction that if

$$\rho_s(z) = \begin{cases} 0, & z > 0 \\ \rho_0, & z < 0 \end{cases},$$

as was the case in Section 3.2, then, in $z < 0$, w will be a potential function (assuming suitable initial conditions). Moreover, by integrating (3.22) across $z = 0$, we find that w is continuous there and, from (3.24), we get that $\bar{p}$ is also continuous, which are the conditions used in deriving the free surface boundary conditions (3.13).

In order to check the validity of (3.25), once again we can systematically non-dimensionalize the equations by writing

$$\rho = \rho_s + \varepsilon\rho_0\hat{\rho}, \qquad p = p_s + \varepsilon p_0\hat{p}, \qquad \mathbf{u} = u_0(\hat{u}, 0, \hat{w}),$$

$x = LX$, $z = LZ$, and $t = \omega_0^{-1}T$. Here, we choose typical values $\rho_0 = \rho_s(0)$ and $p_0 = p_s(0)$, and L and u_0 are, as usual, representative length and velocity scales. Now, the linearized equations (3.21) and (3.22) are obtained from (3.17) and (3.18) as long as $\omega_0 = u_0/\varepsilon L$. Moreover, (3.19) leads to

$$(\rho_s + \varepsilon\rho_0\hat{\rho})\left[\frac{\partial\hat{u}}{\partial T} + \varepsilon\left(\hat{u}\frac{\partial\hat{u}}{\partial X} + \hat{w}\frac{\partial\hat{u}}{\partial Z}\right)\right] = -\frac{\varepsilon^2 p_0}{u_0^2}\frac{\partial\hat{p}}{\partial X}$$

and

$$(\rho_s + \varepsilon\rho_0\hat{\rho})\left[\frac{\partial\hat{w}}{\partial T} + \varepsilon\left(\hat{u}\frac{\partial\hat{w}}{\partial X} + \hat{w}\frac{\partial\hat{w}}{\partial Z}\right)\right] = -\frac{\varepsilon^2 p_0}{u_0^2}\frac{\partial\hat{p}}{\partial Z} - \frac{\varepsilon^2\rho_0 gL}{u_0^2}\hat{\rho}.$$

Hence, in order to retrieve (3.23) and (3.24), we need

$$p_0 \simeq \rho_0 gL \quad \text{and} \quad u_0 \simeq \varepsilon\sqrt{gL}.$$

This example again illustrates the importance of our systematic method. We have chosen the above scales in order to justify the use of (3.25). However, were we modeling sonic boom propagation in the atmosphere, we would be considering wavelengths much shorter than the length scale of the stratification, and this leads to quite a different model, as we will see at the end of this section.

We can extend the theory to disturbances that vary in three dimensions about the same basic stratified equilibrium solution and the equation for w becomes

$$\frac{\partial^2}{\partial t^2}\left(\frac{\partial^2 w}{\partial x^2} + \frac{\partial^2 w}{\partial y^2} + \frac{\partial^2 w}{\partial z^2}\right) = -N^2(z)\left(\left(\frac{\partial^2 w}{\partial x^2} + \frac{\partial^2 w}{\partial y^2}\right) - g^{-1}\frac{\partial^3 w}{\partial z\partial t^2}\right).$$
$$(3.26)$$

We note that the stratification of the fluid destroys any hope of conservation of vorticity. Even in the linear three-dimensional theory, the only vestige that remains is the following argument. Since, from the generalizations of (3.23) and (3.24),

$$\rho_s(z)\frac{\partial\mathbf{u}}{\partial t} = -\nabla\bar{p} - \bar{\rho}g\mathbf{k},$$

we can deduce that

$$\rho_s\frac{\partial}{\partial t}(\nabla\wedge\mathbf{u}) + \rho_s'\mathbf{k}\wedge\frac{\partial\mathbf{u}}{\partial t} = -g\nabla\bar{\rho}\wedge\mathbf{k}$$

and so

$$\mathbf{k}\cdot\frac{\partial\boldsymbol{\omega}}{\partial t} = 0.$$

Hence, the vertical component of the vorticity is conserved in time.

As suggested earlier, it is interesting to note what happens when we combine some aspects of this section with those of Section 3.1 and consider *acoustic waves* in an inhomogeneous compressible atmosphere. Then, we have to revert to the full continuity equation $d\rho/dt + \rho\nabla\cdot\mathbf{u} = 0$. For simplicity, we neglect the effect of gravity, so that $\rho = \rho_s(z)$, but $p_s(z) = $ constant.

The continuity equation linearizes to

$$\frac{\partial \bar{\rho}}{\partial t} + \rho_s \nabla \cdot \mathbf{u} + \mathbf{u} \cdot \nabla \rho_s(z) = 0$$

and the momentum equation is just

$$\rho_s(z) \frac{\partial \mathbf{u}}{\partial t} = -\nabla \bar{p}.$$

We now need to close the system with the energy equation $d/dt(p/\rho^\gamma) = 0$, which linearizes to

$$\frac{1}{p_s} \frac{\partial \bar{p}}{\partial t} = \frac{\gamma}{\rho_s(z)} \left(\frac{\partial \bar{\rho}}{\partial t} + \mathbf{u} \cdot \nabla \rho_s(z) \right).$$

Thus, when we write $\gamma p_s / \rho_s(z) = c_s^2(z)$, we find that the flow is described by a velocity potential ϕ such that $\bar{p} = -\partial \phi / \partial t$, and $\rho_s(z) \mathbf{u} = \nabla \phi$, where

$$\frac{\partial^2 \phi}{\partial t^2} = -\frac{\partial \bar{p}}{\partial t} = \gamma p_s \nabla \left(\frac{1}{\rho_s(z)} \nabla \phi \right)$$

$$= \nabla(c_s^2 \nabla \phi).$$

Note that this result is *not* what we would have obtained by setting $c_0 = c_s(z)$ in (3.6), and although the pressure perturbations satisfy the same equation as ϕ, the density perturbations do not.

3.4 Waves in Rotating Incompressible Flows

It can be shown (see Acheson [5]) that the equations of motion of a constant-density inviscid fluid which is moving with velocity $\mathbf{u}$ relative to a set of axes which are rotating with *constant* angular velocity Ω with respect a fixed inertial frame are

$$\nabla \cdot \mathbf{u} = 0,$$

$$\frac{\partial \mathbf{u}}{\partial t} + (\mathbf{u} \cdot \nabla)\mathbf{u} + 2\Omega \wedge \mathbf{u} + \Omega \wedge (\Omega \wedge \mathbf{r}) = -\frac{1}{\rho} \nabla p. \qquad (3.27)$$

Here, $\mathbf{r}$ is the position vector, in the rotating frame, of the fluid particle whose velocity in that frame is $\mathbf{u}$ and, most importantly, all spatial derivatives are taken relative to the rotating frame. An elementary argument to explain (3.27) is based on the formula that the rate of change of any vector $\mathbf{a}$ with respect to a rotating frame is

$$\frac{d\mathbf{a}}{dt} + \Omega \wedge \mathbf{a}.$$

Hence, the velocity of the particle with position vector $\mathbf{r}$ is

$$\frac{d\mathbf{r}}{dt} + \Omega \wedge \mathbf{r} = \mathbf{u} + \Omega \wedge \mathbf{r}$$

and its acceleration will be

$$\left(\frac{d}{dt} + \boldsymbol{\Omega} \wedge \mathbf{r}\right)(\mathbf{u} + \boldsymbol{\Omega} \wedge \mathbf{r}) = \frac{d\mathbf{u}}{dt} + 2\boldsymbol{\Omega} \wedge \mathbf{u} + \boldsymbol{\Omega} \wedge (\boldsymbol{\Omega} \wedge \mathbf{r}),$$

and, to account for convection, we must interpret $d/dt = \partial/\partial t + \mathbf{u} \cdot \nabla$. This is a plausible but by no means a watertight argument! We can immediately simplify (3.27) since the term $\boldsymbol{\Omega} \wedge (\boldsymbol{\Omega} \wedge \mathbf{r}) = -\nabla(\frac{1}{2}(\boldsymbol{\Omega} \wedge \mathbf{r})^2)$ and, thus, incorporating a centrifugal term in the pressure leads to

$$\frac{\partial \mathbf{u}}{\partial t} + (\mathbf{u} \cdot \nabla)\mathbf{u} + 2\boldsymbol{\Omega} \wedge \mathbf{u} = -\frac{1}{\rho}\nabla p', \tag{3.28}$$

where the *reduced pressure* $p' = p - \frac{1}{2}\rho|\boldsymbol{\Omega} \wedge \mathbf{r}|^2$. Now, a handwaving linearization about an equilibrium state $\mathbf{u} = \mathbf{0}$, $p' = p_0$ leads to

$$\frac{\partial \mathbf{u}}{\partial t} + 2\boldsymbol{\Omega} \wedge \mathbf{u} = -\frac{1}{\rho}\nabla p', \tag{3.29}$$

and a systematic analysis along the lines used in the previous three sections reveals that the nonlinear term in (3.28) can be neglected if the *Rossby number*, Ro, defined as $U_0/L\Omega$, is small. The systematic analysis also shows that the appropriate timescale for this flow is Ω^{-1}. For meteorological flows on the surface of the earth, we might choose $L = 10^3$ km, $U_0 = 10$ ms^{-1}, and, of course, Ω is one revolution per day, so that Ro $\simeq 0.15$. Also, we note that for a steady flow, (3.29) shows that $\mathbf{u} \cdot \nabla p' = 0$; this explains why the wind velocity is parallel to the isobars on which the reduced pressure is constant, as we see daily on weather maps. The term $2\boldsymbol{\Omega} \wedge \mathbf{u}$ in (3.29) is called the *Coriolis term*.

Alas, as in stratified fluids, the flow governed by (3.29) inevitably results in vorticity generation when $\boldsymbol{\Omega} \neq \mathbf{0}$. However, if we take $\boldsymbol{\Omega} = \Omega\mathbf{k}$, it is easy to show from (3.29) that p' and each component of $\mathbf{u}$ all satisfy the equation

$$\frac{\partial^2}{\partial t^2}\left(\frac{\partial^2 \phi}{\partial x^2} + \frac{\partial^2 \phi}{\partial y^2} + \frac{\partial^2 \phi}{\partial z^2}\right) = -4\Omega^2\frac{\partial^2 \phi}{\partial z^2}. \tag{3.30}$$

As stated by Greenspan [6], "the balance between pressure gradient and Coriolis force emerges as the backbone of the entire subject (of rotating flows)." Already we can see the importance of Ω in determining the frequency of oscillatory solutions of (3.30) and the similarities and differences between this model and the inertial wave model given by (3.26).

3.5 Isotropic Electromagnetic and Elastic Waves

Our motivation for now introducing models from the two physically disparate situations of electromagnetics and elasticity is principally to indicate the breadth of applicability of the mathematical methodology that will be described in Chapter 4. Electromagnetism and elasticity are vast subjects, to the modeling of which we cannot hope to do justice here.

Both of these situations have the saving grace of leading to linear models more or less from the start. Maxwell's equations of electromagnetism are deceptively simple, and in free space, they simply state that the electric field $\mathbf{E}$ and the magnetic field $\mathbf{H}$ are related by

$$\nabla \wedge \mathbf{H} = \varepsilon \frac{\partial \mathbf{E}}{\partial t}$$

and

$$\nabla \wedge \mathbf{E} = -\mu \frac{\partial \mathbf{H}}{\partial t}, \tag{3.31}$$

where ε and μ are positive constants[4] and $\nabla \cdot \mathbf{E} = \nabla \cdot \mathbf{H} = 0$. Unfortunately, an explanation of these equations can take many pages, but a simple derivation is described in Coulson and Boyd [7]. For our purposes, the principal result is that all the components of $\mathbf{E}$ and $\mathbf{H}$ satisfy the wave equation

$$\nabla^2 \phi = \frac{1}{c^2} \frac{\partial^2 \phi}{\partial t^2}, \tag{3.32}$$

where $c = 1/\sqrt{\varepsilon \mu}$ is now the speed of light.

The most familiar models for elastic waves are those for the small transverse vibrations of a string or membrane of density ρ under tension T; the displacement simply satisfies the one- or two-dimensional version of (3.32) with $c^2 = T/\rho$. However, the equation for waves in a linear elastic solid looks a little more formidable and can be written as

$$\rho \frac{\partial^2 \mathbf{u}}{\partial t^2} = (\lambda + 2\mu)\nabla(\nabla \cdot \mathbf{u}) - \mu \nabla \wedge (\nabla \wedge \mathbf{u}), \tag{3.33}$$

where $\mathbf{u}$ is the displacement of an element of the material from its equilibrium position and λ and μ are called the Lamé constants[5] of the material. As may be guessed, this equation represents conservation of momentum and has some similarities with the Navier–Stokes equation for viscous flow. The derivation of (3.33) is given by Love [8], where it is explained that the term involving λ represents the stresses that do work in expansion or compression, like the pressure term in (2.7), whereas the terms in μ represent the shear stresses and are analogous to the viscous terms in the Navier–Stokes equation.

Now, we have a more complicated equation, but we can notice at once that if $\mathbf{u} = \nabla \phi$, then (3.33) reduces to

$$\nabla^2 \phi = \frac{1}{c_p^2} \frac{\partial^2 \phi}{\partial t^2}, \tag{3.34}$$

---

[4] They are matrices in an anisotropic medium.
[5] Again, these numbers need to be replaced by matrixes for anisotropic materials.

where $c_p^2 = (\lambda + 2\mu)/\rho$, and if $\mathbf{u} = \nabla \wedge \mathbf{w}$, then the components of $\mathbf{w}$ all satisfy

$$\nabla^2 \phi = \frac{1}{c_s^2} \frac{\partial^2 \phi}{\partial t^2}, \tag{3.35}$$

where $c_s^2 = \mu/\rho$. The key novelty here is that there are two distinct wave speeds, c_p and c_s, and we may anticipate that general solutions of (3.33) will comprise both types of wave.

Since the field equations for $\mathbf{E}$, $\mathbf{H}$, and $\mathbf{u}$ are all vector wave equations, this is a good place to introduce the ideas of *transverse* and *longitudinal* wave motion. In general, waves satisfying vectorial wave equations are *longitudinal* if the vector variable is parallel to the direction of wave motion and *transverse* if it is perpendicular to the direction of the wave. In elasticity, a simple traveling wave solution of (3.34) is $\phi = f(\mathbf{x} \cdot \mathbf{k}_1 - c_p t)$, where $\mathbf{k}_1$ is a fixed unit vector in the wave direction and $\mathbf{u} = \nabla \phi = \mathbf{k}_1 f'(\mathbf{x} \cdot \mathbf{k}_1 - c_p t)$, which is, therefore, an example of a longitudinal wave. The same solution shows that acoustic waves which satisfy (3.6) are always longitudinal since the velocity is in the direction in which the wave travels. On the other hand, if, in (3.35), we take $\mathbf{w} = \mathbf{k}_2 f(\mathbf{x} \cdot \mathbf{k}_1 - c_s t)$, where $\mathbf{k}_2$ is another constant unit vector, then $\mathbf{u} = \nabla f \wedge \mathbf{k}_2 = f'(\mathbf{x} \cdot \mathbf{k}_1 - c_s t) \mathbf{k}_1 \wedge \mathbf{k}_2$ and, hence, this is a transverse wave. Similarly in electromagnetics, there are transverse wave solutions of the form

$$\mathbf{E} = \mathbf{k}_2 f(\mathbf{x} \cdot \mathbf{k}_1 - ct), \qquad \mathbf{H} = \frac{\mathbf{k}_3}{\mu c} f(\mathbf{x} \cdot \mathbf{k}_1 - ct),$$

where $\mathbf{k}_1$, $\mathbf{k}_2$, and $\mathbf{k}_3$ form an orthonormal triad of vectors. In practice, the functions f in all these examples are usually taken to be complex exponentials so that general solutions can be found by Fourier superposition.

The classification of waves into longitudinal and transverse leads to the more general concept of *polarization*[6] in vectorial wave equations. For any such equation with constant coefficients, we may seek solutions of the form

$$\mathbf{u} = f(\mathbf{k} \cdot \mathbf{x} - ct) \mathbf{U},$$

where $\mathbf{k}$ is a unit vector in the direction of the wave and $\mathbf{U}$ is a unit vector which depends on the choice of $\mathbf{k}$. A particular solution of this type is called a *plane polarized* wave, with $\mathbf{U}$ being the *direction* of polarization and $(\mathbf{U}, \mathbf{k})$ defining the *plane* of polarization (assuming it is not a longitudinal wave). As we will see in the next chapter, the general solution can always be written in principle in terms of Fourier integrals as

$$\mathbf{u} = \iiint e^{i(\mathbf{k} \cdot \mathbf{x} - ct)} \mathbf{A}(\mathbf{k}) \, d\mathbf{k};$$

polarized waves correspond to the vector $\mathbf{A}$ being "localized" near a particular vector $\mathbf{k}$.

---

[6] This is not to be confused with *magnetic polarization*, which is an important phenomenon in electromagnetic theory (see Coulson and Boyd [7]).

Exercises

R3.1 (i) Show that, in three dimensions, the linearized equations for acoustic
flow, namely (3.1) and (3.2), are replaced by

$$\frac{\partial \bar{\rho}}{\partial t} + \rho_0 \nabla \cdot \mathbf{u} = 0$$

and

$$\frac{\partial \mathbf{u}}{\partial t} + \frac{1}{\rho_0} \nabla \bar{p} = 0$$

and deduce that $\bar{p}$, $\bar{\rho}$, and $\mathbf{u}$ all satisfy (3.6).

(ii) Suppose now that $\bar{\mathbf{u}} = \mathbf{u} - U\mathbf{i}$ is small, in addition to $\bar{p}$ and $\bar{\rho}$. Show
that the linearized equations are then

$$\left(\frac{\partial}{\partial t} + U \frac{\partial}{\partial x} \right) \bar{\rho} + \rho_0 \nabla \cdot \bar{\mathbf{u}} = 0$$

and

$$\left(\frac{\partial}{\partial t} + U \frac{\partial}{\partial x} \right) \bar{\mathbf{u}} + \frac{1}{\rho_0} \nabla \bar{p} = 0$$

and deduce that $\bar{p}$, $\bar{\rho}$, and $\bar{\mathbf{u}}$ all satisfy

$$\nabla^2 \phi = \frac{1}{c_0^2} \left(\frac{\partial}{\partial t} + U \frac{\partial}{\partial x} \right)^2 \phi.$$

Show that this reduces to (3.6) if we change to moving axes (X, y),
where $X = x - Ut$ and we assume that U is constant.

R3.2 Gas is contained in a box $0 < x < L$, $0 < y < b$, $0 < z < c$. Show that
acoustic oscillations satisfying (3.6) are possible in which ϕ is proportional
to $\cos \omega t$ if

$$\omega^2 = \pi^2 c_0^2 \left(\frac{l^2}{L^2} + \frac{m^2}{b^2} + \frac{n^2}{c^2} \right),$$

where l, m, and n are integers.

Show also that if just one face of the box is subject to small-amplitude
oscillations so that

$$\frac{\partial \phi}{\partial x} = a \cos \omega t$$

on $x = 0$, then, in general, a possible solution is

$$\phi = \frac{-ac_0 \cos[\omega(L - x)/c_0]}{\omega \sin \omega L / c_0} \cos \omega t.$$

For what values of ω is this solution inadmissible?

Show that if $L = \infty$ and

$$\frac{\partial \phi}{\partial x} = a \cos \frac{m\pi y}{b} \cos \frac{n\pi z}{c} \cos \omega t$$

on $x = 0$, then if $m^2/b^2 + n^2/c^2 > \omega^2/c_0^2\pi^2$, there are solutions of the form

$$\phi = -\frac{a}{\lambda}\cos\frac{m\pi y}{b}\cos\frac{n\pi z}{c}e^{-\lambda x}\cos\omega t,$$

where $\lambda^2 = m^2\pi^2/b^2 + n^2\pi^2/c^2 - \omega^2/c_0^2$. Show further that if $m^2/b^2 + n^2/c^2 < \omega^2/c_0^2\pi^2$, then

$$\phi = -\frac{a}{\mu}\cos\frac{m\pi y}{b}\cos\frac{n\pi z}{c}\sin(\omega t - \mu x),$$

where $\mu^2 = \omega^2/c_0^2 - m^2\pi^2/b^2 - n^2\pi^2/c^2$. (This problem of a *wavemaker* will be considered further in Chapter 4.)

3.3 Show that if $\phi(r,t)$ is the velocity potential for a spherically symmetric acoustic wave, where r is the polar coordinate measured from the origin, then

$$\frac{\partial^2\phi}{\partial r^2} + \frac{2}{r}\frac{\partial\phi}{\partial r} = \frac{1}{c_0^2}\frac{\partial^2\phi}{\partial t^2}.$$

Deduce that $r\phi$ satisfies the one-dimensional wave equation.

Acoustic waves in an infinite gas are driven by a sphere which starts oscillating at $t = 0$ so that its radius is given by $r = a(1 + \varepsilon\cos\omega t)$, where $\varepsilon \ll 1$. Show that the appropriate boundary condition for acoustic waves in $r > a$ is

$$\frac{\partial\phi}{\partial r} = -a\varepsilon\omega\sin\omega t \quad\text{on } r = a.$$

Show that for $t > 0$, the velocity potential ϕ is given by

$$\phi = \frac{1}{r}\left[\frac{a^3\varepsilon\omega^2 c_0^2}{c_0^2 + a^2\omega^2}\right]\left[a\omega\cos\frac{\omega}{c_0}(r - a - c_0 t)\right.$$
$$\left. + c_0\sin\frac{\omega}{c_0}(r - a - c_0 t) - a\omega e^{(1/a)(r-a-c_0 t)}\right]$$

for $a < r < a + c_0 t$.

R3.4 Show that for small-amplitude waves on an incompressible stream in which $\mathbf{u} = \nabla(Ux + \phi)$, where ϕ and the elevation η are small, the linearized versions of the boundary conditions (3.10) and (3.11) are

$$\frac{\partial\phi}{\partial z} = \frac{\partial\eta}{\partial t} + U\frac{\partial\eta}{\partial x}$$

and

$$\frac{\partial\phi}{\partial t} + U\frac{\partial\phi}{\partial x} + g\eta = 0$$

on $z = 0$. If $\eta = a\cos(kx - \omega t)$, show that a solution of (3.8) satisfying $\partial\phi/\partial z = 0$ on $z = -h$ and the above boundary conditions is

$$\phi = \frac{a(\omega - Uk)\cosh k(z + h)\sin(kx - \omega t)}{k\sinh kh},$$

providing $(\omega - Uk)^2 = gk \tanh kh$. Deduce that ω and k can only both be real if $g > 0$.

Show that a solution for steady waves (with $\omega = 0$) is only possible if $U^2 < gh$. Show also that if $U = 0$, then, as $h \to \infty$, $\omega^2 \to gk$.

3.5 Show that three-dimensional Stokes waves on the surface of a running stream of depth h can be found where the surface elevation is

$$\eta = a\cos(k_1 x + k_2 y - \omega t)$$

and the velocity potential is

$$\phi = Ux + b\cosh\left[\sqrt{k_1^2 + k_2^2}(z + h)\right]\sin(k_1 x + k_2 y - \omega t),$$

provided

$$(Uk_1 - \omega)^2 = g\sqrt{k_1^2 + k_2^2}\tanh\sqrt{k_1^2 + k_2^2}\,h$$

and

$$b = \frac{a(\omega - Uk_1)}{\sqrt{k_1^2 + k_2^2}\sinh\sqrt{k_1^2 + k_2^2}\,h}.$$

3.6 Small-amplitude waves propagate on the interface $z = 0$, which separates liquid of density ρ_1 in $z > 0$ from liquid of density ρ_2 in $z < 0$. The upper liquid is streaming with uniform velocity U in the x direction and the lower fluid is at rest. If variables in the upper and lower liquids are denoted by suffices 1 and 2, respectively, and $z = \eta$ is the elevation of the interface, show that the model (3.12)–(3.14) generalizes to

$$\nabla^2\phi_1 = 0 \quad \text{in } z > 0, \qquad \nabla^2\phi_2 = 0 \quad \text{in } z < 0,$$

with

$$\frac{\partial\phi_1}{\partial z} = \frac{\partial\eta}{\partial t} + U\frac{\partial\eta}{\partial x}, \qquad \frac{\partial\phi_2}{\partial z} = \frac{\partial\eta}{\partial t},$$

and

$$\rho_1\left(\frac{\partial\phi_1}{\partial t} + U\frac{\partial\phi_1}{\partial x} + g\eta\right) = \rho_2\left(\frac{\partial\phi_2}{\partial t} + g\eta\right)$$

on $z = 0$. Show that waves for which $\eta = a\cos(kx - \omega t)$, with $k > 0$, are possible provided

$$\rho_1((\omega - Uk)^2 + gk) = \rho_2(gk - \omega^2).$$

Deduce that when $U = 0$, with $g > 0$, ω and k can only both be real if $\rho_2 \geq \rho_1$. Show also that ω and k cannot both be real when $g = 0$ and $U \neq 0$.

3.7 From (2.6) and (2.7), show that waves propagating in a vertical direction in an inhomogeneous atmosphere satisfy

$$\frac{\partial \rho}{\partial t} + \frac{\partial}{\partial z}(\rho w) = 0,$$

$$\rho\left(\frac{\partial w}{\partial t} + w\frac{\partial w}{\partial z}\right) = -\frac{\partial p}{\partial z} - g\rho,$$

and

$$\left(\frac{\partial}{\partial t} + w\frac{\partial}{\partial z}\right)\left(\frac{p}{\rho^\gamma}\right) = 0.$$

Show that, in equilibrium, $\rho = \rho_s(z)$ and $p = p_s(z)$ satisfy (3.20). For acoustic waves, the variables w, $\bar{\rho} = \rho - \rho_s$, and $\bar{p} = p - p_s$ are all small. Show that the linearized equations satisfied by these variables are

$$\frac{\partial \bar{\rho}}{\partial t} + \rho_s\frac{\partial w}{\partial z} + \rho'_s w = 0,$$

$$\rho_s\frac{\partial w}{\partial t} = -\frac{\partial \bar{p}}{\partial z} - g\bar{\rho}$$

and

$$\frac{\partial \bar{p}}{\partial t} - c_s^2\frac{\partial \bar{\rho}}{\partial t} + w(p'_s - c_s^2\rho'_s) = 0,$$

where $c_s^2 = \gamma p_s/\rho_s$. If we assume that gravity is negligible, show that p_s is constant and deduce that

$$\frac{\partial^2 \bar{p}}{\partial t^2} = \frac{\partial}{\partial z}\left(c_s^2\frac{\partial \bar{p}}{\partial z}\right).$$

3.8 A component of a heat exchanger consists of a uniform tube along the x axis which contains gas and whose walls transmit heat to the gas at a rate $\partial Q(x,t)/\partial t$ per unit length. When $Q = 0$, the gas has constant speed U, density ρ_0, and pressure p_0. Show that for small heat addition, the pressure, density, and velocity perturbations satisfy the equations

$$\frac{\partial \bar{\rho}}{\partial t} + \rho_0\frac{\partial \bar{u}}{\partial x} + U\frac{\partial \bar{\rho}}{\partial x} = 0,$$

$$\rho_0\frac{\partial \bar{u}}{\partial t} + \rho_0 U\frac{\partial \bar{u}}{\partial x} = -\frac{\partial \bar{p}}{\partial x}$$

and

$$\rho_0 c_v\left(\frac{\partial \bar{T}}{\partial t} + U\frac{\partial \bar{T}}{\partial x}\right) = \frac{p_0}{\rho_0}\left(\frac{\partial \bar{\rho}}{\partial t} + U\frac{\partial \bar{\rho}}{\partial x}\right) + \frac{\partial Q}{\partial t},$$

where $\bar{T} = \bar{p}/\rho_0 R - p_0\bar{\rho}/\rho_0^2 R$. Deduce that

$$\left(\frac{\partial}{\partial t} + U\frac{\partial}{\partial x}\right)\left(\frac{\partial^2 \bar{p}}{\partial t^2} + 2U\frac{\partial \bar{\rho}}{\partial x \partial t} + (U^2 - c_0^2)\frac{\partial^2 \bar{\rho}}{\partial x^2}\right) = (\gamma - 1)\frac{\partial^3 Q}{\partial x^2 \partial t}.$$

3.9 Suppose that the Rossby number in an incompressible rotating fluid is small so that (3.29) holds with respect to a frame rotating with angular velocity $\mathbf{\Omega}$. Show that if $\mathbf{\Omega} = (0, 0, \Omega)$, then, in steady flow, $\partial \mathbf{u}/\partial z = \mathbf{0}$ (this is the *Taylor–Proudman theorem*).

Now suppose that fluid fills a sealed cylindrical container which is rotating with a small Rossby number. The ends of the cylinder are flat and perpendicular to the axis of rotation save for a small finite bump on one end which protrudes into the fluid.

Show that an observer rotating with the cylinder will see a two-dimensional flow perpendicular to the axis of rotation in which a "pillar" of fluid above the bump is at rest (this pillar is called a *Taylor column*).

R3.10 A long tube containing gas at rest lies along the x axis. In $x < 0$, the gas has density ρ_1 and sound speed c_1, whereas in $x > 0$, the gas has density ρ_2 and sound speed c_2. An acoustic wave described by $\phi = a \sin k(c_1 t - x)$ is incident from the region $x < 0$. Show that at $x = 0$, $\partial\phi/\partial x$ and $\rho(\partial\phi/\partial t)$ are continuous and deduce that the reflected and transmitted waves have amplitude aR and aT, respectively, where

$$R = \left| \frac{\rho_2 c_2 - \rho_1 c_1}{\rho_1 c_1 + \rho_2 c_2} \right| \quad \text{and} \quad T = \left| \frac{2\rho_1 c_2}{\rho_2 c_2 + \rho_1 c_1} \right|.$$

(R is the *reflection coefficient* and T is the *transmission coefficient*.)

This illustrates the idea of the *impedance* of a boundary, which is a generic expression used to describe the qualitative response of an inhomogeneity to an incoming wave. In this case we can see that if $\rho_1 c_1 = \rho_2 c_2$, there is no reflected wave. Hence, when $\rho_1 c_1 - \rho_2 c_2$ is suitably small, we say that the boundary has low impedance, whereas if ρ_1 or c_2 is suitably small, the transmission is weak and it has high impedance.

*3.11 In this Exercise, $[\ ]$ is used to denote the size of a discontinuous jump in a variable.

(i) The vector $\mathbf{a}$ satisfies $\nabla \cdot \mathbf{a} = 0$ and changes rapidly from one side of a surface S to the other. By integrating over a "pillbox" straddling an area Σ of S with normal $\mathbf{n}$ and then shrinking the pillbox to zero, show that

$$[\mathbf{a} \cdot \mathbf{n}]_-^+ = 0$$

in the limit when $\mathbf{a}$ has a jump discontinuity across S.

(ii) The matrix $A = (A_{ij})$ satisfies the equation $\partial A_{ij}/\partial x_j = 0$. Show that if A has a jump discontinuity across S, then

$$[A\mathbf{n}]_-^+ = \mathbf{0}.$$

(iii) Show that $(\nabla \wedge \mathbf{b})_i = -\partial A_{ij}/\partial x_j$ if

$$A = \begin{pmatrix} 0 & -b_3 & b_2 \\ b_3 & 0 & -b_1 \\ -b_2 & b_1 & 0 \end{pmatrix}.$$

Hence, deduce that if $\nabla \wedge \mathbf{b} = \mathbf{0}$ and $\mathbf{b}$ has a jump discontinuity across S, then

$$[\mathbf{b} \wedge \mathbf{n}]_-^+ = \mathbf{0}.$$

(iv) When ε and μ are spatially dependent, Maxwell's equations can be written as

$$\nabla \cdot (\varepsilon \mathbf{E}) = 0,$$
$$\nabla \cdot (\mu \mathbf{H}) = 0,$$
$$\mu \frac{\partial \mathbf{H}}{\partial t} = -\nabla \wedge \mathbf{E}$$

and

$$\varepsilon \frac{\partial \mathbf{E}}{\partial t} = \nabla \wedge \mathbf{H}.$$

Show that across a surface on which $\mathbf{E}$ and $\mathbf{H}$ have jump discontinuities,

$$[\varepsilon \mathbf{E} \cdot \mathbf{n}] = 0, \qquad [\mu \mathbf{H} \cdot \mathbf{n}] = 0,$$
$$[\mathbf{E} \wedge \mathbf{n}] = \mathbf{0}, \qquad [\mathbf{H} \wedge \mathbf{n}] = \mathbf{0}.$$

Note that if conducting material is present, Maxwell's equations have to be modified to allow for current flow. Hence, these jump conditions may not be appropriate at the boundary of a conductor.

*3.12 By writing $\mathbf{u} = \nabla \phi + \nabla \wedge \mathbf{\Psi}$, where $\nabla \cdot \mathbf{\Psi} = 0$, show that (3.33) for elastic waves can be satisfied if

$$\frac{\partial^2 \phi}{\partial t^2} = c_p^2 \nabla^2 \phi$$

and

$$\frac{\partial^2 \mathbf{\Psi}}{\partial t^2} = c_s^2 \nabla^2 \mathbf{\Psi},$$

where $c_p = (\lambda + 2\mu/\rho)^{1/2}$ and $c_s = (\mu/\rho)^{1/2}$.

Consider waves traveling in the x direction in a semi-infinite elastic solid $z \geq 0$. Given that $\phi = \phi(x, z, t)$, $\mathbf{\Psi} = (0, -\psi(x, z, t), 0)$, and that the boundary condition on the free surface $z = 0$ is $\sigma_{i3} = 0$ where, with $(x, z) = (x_1, x_3)$,

$$\sigma_{ij} = \delta_{ij} \lambda \nabla \cdot \mathbf{u} + \mu \left(\frac{\partial u_i}{\partial x_j} + \frac{\partial u_j}{\partial x_i} \right),$$

show that on $z = 0$,

$$\lambda \left(\frac{\partial^2 \phi}{\partial x^2} + \frac{\partial^2 \phi}{\partial z^2} \right) + 2\mu \left(\frac{\partial^2 \phi}{\partial z^2} - \frac{\partial^2 \psi}{\partial x \partial z} \right) = 0$$

and

$$2 \frac{\partial^2 \phi}{\partial x \partial z} - \frac{\partial^2 \psi}{\partial x^2} + \frac{\partial^2 \psi}{\partial z^2} = 0.$$

3.13 A sound source with frequency ω moves along a tube with speed V for time $t > 0$. You are given that the velocity potential satisfies

$$\frac{\partial^2 \phi}{\partial x^2} = \frac{1}{c^2} \frac{\partial^2 \phi}{\partial t^2}, \quad x \neq Vt,$$

where ϕ is continuous at the source $x = Vt$ and the velocity jump across $x = Vt$ is $\cos \omega t$. If the gas is initially at rest, show that if $V < c$,

$$\frac{2\omega c}{c^2 - V^2} \phi = \begin{cases} -\sin \dfrac{(ct - x)\omega}{c - V}, & Vt < x < ct \\[2ex] -\sin \dfrac{(ct + x)\omega}{c + V}, & -ct < x < Vt. \end{cases}$$

Deduce that the Doppler frequency shift between observers just ahead of and behind the source is

$$\omega c \left(\frac{1}{c - V} - \frac{1}{c + V} \right) = \frac{2\omega c V}{c^2 - V^2}.$$

4

Theories for Linear Waves

Looking back at the models derived in the last chapter, we see that they almost all comprise linear partial differential equations in time and at least one space variable, together with linear boundary conditions. The only exception to this rule is the model describing Stokes waves in Section 3.2, where the time derivatives only occur in the boundary condition. Importantly, in most of the equations, many of the terms have constant coefficients. We therefore start this chapter by reviewing the mathematical methodologies that are available for the analysis of such models.

4.1 Wave Equations and Hyperbolicity

It is well known that systems of linear partial differential equations can be classified into a hierarchy which has "hyperbolic" models at one end and "elliptic" models at the other (Ockendon et al. [9]). Hyperbolic models are probably the best understood and for such a system the *Cauchy problem*, in which appropriate data are prescribed at some initial time, is, in general, a well-posed problem. Moreover, much is known about how the solution depends on the data via *regions of influence* and *domains of dependence* in the space of the independent variables. These concepts depend crucially on the fact that the *characteristics* (in two dimensions) or the *characteristic manifolds* (in three or more dimensions) are real for a hyperbolic system. However, this information does not necessarily give us any detailed knowledge of the solution itself, let alone an explicit analytic solution.

The situation can be illustrated with reference to the wave equation (3.4) in one space dimension,

$$\frac{\partial^2 \phi}{\partial x^2} = \frac{1}{c^2} \frac{\partial^2 \phi}{\partial t^2};$$

(4.1)

for most of this chapter, we drop the suffix zero from c_0 for convenience. It is known (see Ockendon et al. [9]) that the quasilinear partial differential equation

$$A\frac{\partial^2 \phi}{\partial x^2} + 2B\frac{\partial^2 \phi}{\partial x \partial t} + C\frac{\partial^2 \phi}{\partial t^2} = D,$$

where A, B, and C depend only on x and t, is hyperbolic if $B^2 > AC$, and its characteristics are given by

$$C\left(\frac{dx}{dt}\right)^2 - 2B\frac{dx}{dt} + A = 0.$$

Hence we can see at once that (4.1) is hyperbolic and its characteristics are the lines $x \pm ct = $ constant. Moreover, if *Cauchy data*

$$\phi(x,0) = f(x) \quad \text{and} \quad \frac{\partial \phi}{\partial t}(x,0) = g(x) \tag{4.2}$$

are prescribed at $t = 0$ for all x, where, additionally, f and g are only non-zero in a finite interval $a < x < b$, then it is only possible for the solution to be non-zero in $a - ct < x < b + ct$. This is the *region of influence* of the interval (a, b). Even without the general theory, these results follow directly from *D'Alembert's solution*

$$\phi = \frac{1}{2}[f(x-ct) + f(x+ct)] + \frac{1}{2c}\int_{x-ct}^{x+ct} g(s)\,ds. \tag{4.3}$$

Yet another way to look at the solution of (4.1) is by "factorizing" the differential operators and writing the equation in the form

$$\left(\frac{\partial}{\partial x} \mp \frac{1}{c}\frac{\partial}{\partial t}\right)\left(\frac{\partial \phi}{\partial x} \pm \frac{1}{c}\frac{\partial \phi}{\partial t}\right) = 0,$$

so that it follows that $\partial \phi/\partial x \pm (1/c)(\partial \phi/\partial t)$ is constant on the lines $x \pm ct = $ constant. Hence $\phi = F(x - ct) + G(x + ct)$ for arbitrary functions F, G.

We note that by considering Cauchy data with "compact support," (i.e., data that are only non-zero on a finite interval of the x axis), we have found solutions that are not analytic everywhere; hence, we have run the risk of ending up with a solution which is not differentiable enough for (4.1) to make sense. We will return to this restriction in Chapter 6 but will not let this interrupt the discussion for the moment.

It is sad but ineluctible that (4.1) is almost the only model from Chapter 3 whose general solution can be written down explicitly, as in (4.3). One other such case is that of sound waves with spherical symmetry when (3.6) reduces to

$$\frac{\partial^2 \phi}{\partial r^2} + \frac{2}{r}\frac{\partial \phi}{\partial r} = \frac{1}{c^2}\frac{\partial^2 \phi}{\partial t^2}, \tag{4.4}$$

where r is the spherical polar coordinate. At first sight, this looks worse than (4.1) because the extra term does not have a constant coefficient, but writing $r\phi = \Phi$ leads to (4.1) for Φ, so that the general solution is

$$\phi = \frac{1}{r}[F(r+ct) + G(r-ct)]. \tag{4.5}$$

We thus see that acoustic waves in one dimension and in three dimensions are closely related. However, things are not so simple in two dimensions. Writing (3.6) in cylindrical polar coordinates and assuming that the flow is axisymmetric leads to

$$\frac{\partial^2 \phi}{\partial r^2} + \frac{1}{r}\frac{\partial \phi}{\partial r} = \frac{1}{c^2}\frac{\partial^2 \phi}{\partial t^2},$$

and here the term $(1/r)(\partial\phi/\partial r)$ really does make life harder. We will return to this equation and to the fascinating question of how waves depend on the dimensionality of the space within which they are propagating in Section 4.8.

"Wave equations" such as (3.6), (3.25), and (3.30) of Chapter 3 act as a marvellous springboard for a mathematical treatment of wave motion and might provide a basis for the statement that "hyperbolic equations are wave equations". However, surface gravity waves are described by an *elliptic* partial differential equation so how is it that the boundary conditions can allow wave solutions? Also, how can we reconcile hyperbolicity with the observation that if we seek acoustic waves that vary harmonically in time in three dimensions by writing $\phi = \mathrm{Rl}(\Phi(x,y,z)e^{-i\omega t})$, then we are left with the *elliptic* equation $\nabla^2\Phi + (\omega^2/c^2)\Phi = 0$?

These questions suggest that we need a more general idea than that of hyperbolicity if we are to encompass many of the waves that occur in nature.

4.2 Fourier Series, Eigenvalues, and Resonance

Fourier analysis is one of the most powerful methods for the analysis of linear equations, especially ones which have constant coefficients. Irrespective of whether the partial differential equation is hyperbolic, elliptic, or parabolic, such an equation will have solutions that can be obtained by the method of separation of variables, which leads to solutions that are products of exponential functions (with either real or imaginary argument).[1] By summing solutions of this type, it is possible to generate quite general explicit solutions which are often more convenient even than an exact representation such as D'Alembert's solution (4.3) of the one-dimensional wave equation.

If we consider acoustic waves or waves on a finite string satisfying (4.1) and with boundary conditions $\phi = 0$ at $x = 0$ and $x = L$, it is easy to separate the variables and appeal to the theory of Fourier series in order to obtain

$$\phi = \sum_{n=1}^{\infty} \left(a_n \cos\frac{n\pi ct}{L} + b_n \sin\frac{n\pi ct}{L} \right) \sin\frac{n\pi x}{L}. \tag{4.6}$$

For this finite domain, this form of solution is much easier to deal with than the D'Alembert solution (4.3) which will involve an infinite series of reflecting

---

[1] Separation of variables can sometimes be applied to variable-coefficient equations, as will be seen later.

waves. For a Cauchy problem, where the initial values are given by (4.2), the coefficients a_n and b_n can easily be found to be

$$a_n = \frac{2}{L} \int_0^L f(x) \sin \frac{n\pi x}{L} \, dx \tag{4.7a}$$

and

$$b_n = \frac{2}{n\pi c} \int_0^L g(x) \sin \frac{n\pi x}{L} \, dx. \tag{4.7b}$$

This analysis requires that the functions f and g satisfy certain smoothness conditions and the series (4.6) will only converge at points where ϕ is continuous.

The solution (4.6) for one-dimensional waves in a closed container consists of a sum of "eigenmodes" or "normal modes," which can be thought of as the infinite-dimensional generalization of the normal modes encountered when considering small oscillations in classical mechanics. These oscillations are perpetual motions (assuming that there is no damping) which can exist in the absence of any long-term forcing; they just need to be initiated by some given non-zero initial conditions. If the wave is forced by persistent non-zero boundary conditions at $x = 0$ and $x = L$, we can still use this form of the solution but we will need to add in a "particular solution" of (4.1) which satisfies the forcing condition and then the calculation for a_n and b_n will be different. For example, suppose that we wish to solve (4.1) given

$$\phi = 0 \quad \text{on } x = 0 \qquad \text{and} \qquad \phi = \cos \omega t \quad \text{on } x = L,$$

in addition to the usual initial conditions (4.2). Then, we can write the solution as

$$\phi = \frac{\sin(\omega x/c)(\cos \omega t)}{\sin(\omega L/c)} + \sum_{n=1}^{\infty} \left(a_n \cos \frac{n\pi ct}{L} + b_n \sin \frac{n\pi ct}{L} \right) \sin \frac{n\pi x}{L} \tag{4.8}$$

as long as $\omega L/c\pi$ is not in integer, and then, applying the initial conditions, we obtain

$$a_n = \frac{2}{L} \int_0^L \left(f(x) - \frac{\sin(\omega x/c)}{\sin(\omega L/c)} \right) \sin \frac{n\pi x}{L} \, dx$$

and b_n is as given in (4.7b).

We make the important remark that the Fourier series representation (4.6) is not chosen simply because Fourier series are a convenient and familiar way of representing mathematical functions on a finite interval; the form of the "modes" defined by the terms in (4.6) are a direct result of separating the variables in (4.1) and applying the zero (homogeneous) boundary conditions at $x = 0$ and $x = L$. If non-trigonometric "eigenfunctions" had emerged as a result of separation of variables, it would have been appropriate to employ a "generalized" Fourier series expansion in which these eigenfunctions were used as a basis (see Exercise 4.2). This procedure is often needed for the construction of normal modes in more general wave models.

Normal modes, whether in terms of trigonometric functions or not, are of great practical importance because of the phenomenon of *resonance*. We met this idea in Exercise 3.2 and a second example is the revelation in (4.8) that the periodic forced solution will not exist if $\omega = n\pi c/L$ for some integer n. In crude terms, resonance is the surprisingly large-amplitude response that occurs when the boundary forcing is at one of the "natural frequencies" (or "normal frequencies") of the unforced system.

We now consider how the idea of Fourier analysis can be used to help us to understand resonance in a more general situation. Intuitively, we expect a decomposition into eigenmodes to be possible for any linear undamped unforced wave model in a finite domain. To see this mathematically, we represent the wave model by

$$\mathcal{L}\phi = \frac{1}{c_0^2}\frac{\partial^2 \phi}{\partial t^2} \quad \text{in } D \tag{4.9}$$

with

$$\mathcal{L}\phi = 0 \quad \text{on } \partial D,$$

where $\mathcal{L}$ is some linear spatial differential operator. We then seek waves of frequency ω by writing[2]

$$\phi = \mathrm{Rl}(\Phi e^{-i\omega t}). \tag{4.10}$$

This leads to the problem

$$\mathcal{L}\Phi + \frac{\omega^2}{c_0^2}\Phi = 0 \quad \text{in } D \tag{4.11}$$

with

$$\Phi = 0 \quad \text{on } \partial D,$$

which is an eigenvalue problem where the eigenvalues ω^2/c_0^2 determine the natural frequencies of the system (4.9). There is an enormous literature on such problems and the dependence of the eigenvalues on the operator $\mathcal{L}$ and the geometry of D. We will not discuss this further here except to say that under "nice" conditions and when there is no damping, the possible values of ω will be discrete, real, and positive in any closed *resonator* or finite domain D. Typically, the discrete numbers ω will "grow linearly" so that if, say, in a one-dimensional problem, the eigenvalues ω_n are arranged in increasing order of magnitude, then $\omega_n = O(n)$ as $n \to \infty$. In particular, for the one-dimensional acoustic oscillator in $0 < x < L$, we see from (4.6) that $\omega_n = n\pi c_0/L$, and for the three-dimensional version for waves in a rectangular box with sides of length a, b, and c (Exercise 3.2),

$$\omega_{lmn} = \pi c_0 \left[\frac{l^2}{a^2} + \frac{m^2}{b^2} + \frac{n^2}{c^2}\right]^{1/2}, \tag{4.12}$$

---

[2] The negative sign is taken in the exponent for reasons that will become apparent later.

where l, m, and n are all integers.

Armed with this information about eigenvalues, we can quickly encapsulate the phenomenon of resonance in mathematical terms. Suppose that an acoustic resonator is forced to oscillate by having part or all of its boundary moving with frequency ω. If we seek a periodic solution to (3.6) in the form (4.10), we find that

$$\nabla^2 \Phi + \frac{\omega^2}{c^2} \Phi = 0, \qquad (4.13)$$

with Φ prescribed and non-zero on the boundary ∂D. By the Fredholm alternative (see Ockendon et al. [9]), we can assert that this problem will have no solution whenever ω/c is one of the eigenvalues of the solution of (4.13) with zero boundary conditions. Thus, the phenomenon of resonance is simply a manifestation of the Fredholm alternative.

From a different viewpoint, if the resonator is at rest and we start to drive it periodically at one of the eigenfrequencies, we would find that the response grows linearly with time, thereby destroying any possibility of an eventual periodic response. In practice, this unlimited growth is usually mitigated by the effects of some damping in the system, as illustrated in Exercise 4.3. However, in certain cases, the amplitude of the response may become so large as to invalidate the asumptions that were built into the linear approximation. Then, as we will see in Chapter 5, we have to reconsider the perturbation arguments we used in Chapter 3 so as to bring nonlinear terms into play.

We have only defined normal modes and resonances when the wave motion is confined in a closed container. Quite a different situation applies when the motion occurs in a region that extends to infinity in all directions. If we excite such a system by a *transient* localized forcing, we expect all of the energy to propagate to infinity, as in the solution (4.5) with $f=0$, and the motion will eventually die away even if there is no damping. If, however, a *periodic* forcing is maintained, then we will again be able to look for solutions of the form $\phi = \mathrm{Rl}(e^{-i\omega t} \Phi)$, with Φ satisfying (4.13). In this case, the conditions to be imposed at infinity are less obvious and we will return to such problems in Section 4.5.1.

An interesting configuration that is halfway between a bounded "interior" problem and an infinite "exterior" problem is found in a *waveguide*. This is a device in which waves are directed to propagate in a semi-infinite channel; the reflection from the walls of the channel allows the propagation to be unattenuated, unlike the spherically symmetric wave given by (4.5). We can understand this most easily by solving the problem for two-dimensional acoustic waves in a channel with fixed walls at $y = 0$ and $y = b$. The velocity potential ϕ will satisfy

$$\frac{\partial^2 \phi}{\partial x^2} + \frac{\partial^2 \phi}{\partial y^2} = \frac{1}{c^2} \frac{\partial^2 \phi}{\partial t^2}$$

with $\partial\phi/\partial y = 0$ on $y = 0$ and $y = b$; thus, writing $\phi = \mathrm{Rl}(\Phi e^{-i\omega t})$ now leads to solutions of the form

$$\Phi = \cos\frac{n\pi y}{b}(Ae^{ikx} + Be^{-ikx}), \qquad (4.14)$$

where $k^2 = \omega^2/c^2 - n^2\pi^2/b^2$ and n is any integer.[3]

From this, we can immediately discern the crucial attribute of wave guides; this is the fact that the wave can only propagate in the x direction if ω exceeds the so-called *"cut-off"* frequency $c\pi/b$; if this is not the case, k will not be real and (4.14) will not represent a propagating wave.

We conclude this section with one piece of jargon. So far, all our linear wave models have been posed as *evolution problems* in which time appears as an independent variable and we are thus in the "time domain." However, the representations in this section in terms of eigenmodes have inevitably led us to equations like (4.11) or (4.13), in which the only independent variables are spatial and it is assumed that all time variations are harmonic with frequency ω. Such problems are said to be posed in the *frequency domain* and will be considered in more detail in Section 4.5. However, before going down this route, we turn our attention to the Fourier representation of problems in infinite domains.

4.3 Fourier Integrals and the Method of Stationary Phase

In order to represent solutions of models for waves propagating in infinite domains, we start by indicating how the theory for Fourier series can be extended to apply to non-periodic functions. This leads us to *Fourier transform* theory.

Noting that a general, sufficiently smooth, $2l$ periodic function $f(x)$ has the Fourier series

$$f(x) = \frac{a_0}{2} + \sum_1^\infty a_n\cos\frac{n\pi x}{l} + b_n\sin\frac{n\pi x}{l}, \qquad (4.15)$$

where

$$a_n + ib_n = \frac{1}{l}\int_{-l}^{l} f(x)\left(\cos\frac{n\pi x}{l} + i\sin\frac{n\pi x}{l}\right)dx,$$

we can see at once that an alternative formulation of (4.15) is

$$f(x) = \sum_{-\infty}^{\infty} c_n e^{-in\pi x/l}, \quad c_n = \frac{1}{2l}\int_{-l}^{l} f(x)e^{in\pi x/l}\,dx,$$

---

[3] The reason for the minus sign in (4.10) is now apparent, because it means the first term in (4.14) represents waves propagating in the positive x direction.

where $c_n = \frac{1}{2}(a_n + ib_n)$, $a_n = a_{-n}$, and $b_n = -b_{-n}$. This gives us the vital clue as to how to deal with *non-periodic* functions. All we need to do formally is to put $n\pi/l = k$ and let $l \to \infty$ to obtain the Fourier integral transform formulas

$$\bar{f}(k) = \int_{-\infty}^{\infty} f(x)e^{ikx}\,dx \qquad (4.16)$$

and

$$f(x) = \frac{1}{2\pi}\int_{-\infty}^{\infty} \bar{f}(k)e^{-ikx}\,dk. \qquad (4.17)$$

Of course, this leaves open important questions of convergence which we will not address in general terms; when these questions loom large, we will deal with them on a case-by-case basis.

A trivial illustration of this method is the solution of the one-dimensional wave equation (4.1) on $-\infty < x < \infty$. By multiplying (4.1) by e^{ikx} and integrating from $x = -\infty$ to $x = \infty$, we see at once that the Fourier transform of ϕ is $\bar{A}(k)e^{ikct} + \bar{B}(k)e^{-ikct}$, where $\bar{A}$ and $\bar{B}$ are general functions, and hence

$$\phi = \frac{1}{2\pi}\int_{-\infty}^{\infty} \bar{A}(k)e^{-ik(x-ct)}\,dk + \frac{1}{2\pi}\int_{-\infty}^{\infty} \bar{B}(k)e^{-ik(x+ct)}\,dk.$$

This is a solution containing waves corresponding to all values of k, which can be seen to be the sum of two traveling waves

$$\phi = A(x - ct) + B(x + ct),$$

as in (4.3). In wave propagation, the Fourier transform variable k is usually called the *wavenumber* and solutions of (4.1) that are proportional to $e^{\pm ikx}$ have wavelength $2\pi/k$.

The solution of the general initial value problem for two-dimensional surface gravity waves is less trivial. Suppose that we consider waves on water of infinite depth which is initially at rest with a surface elevation $z = \eta_0(x)$. Then, we can take the Fourier transform of (3.12)–(3.14) by writing

$$\bar{\phi}(k, z, t) = \int_{-\infty}^{\infty} \phi(x, z, t)e^{ikx}\,dx,$$

$$\bar{\eta}(k, t) = \int_{-\infty}^{\infty} \eta(x, t)e^{ikx}\,dx,$$

and assuming that $|\phi|$ and $|\nabla\phi| \to 0$ as $|x| \to \infty$, we get

$$\frac{\partial^2 \bar{\phi}}{\partial z^2} - k^2 \bar{\phi} = 0,$$

with

$$\frac{\partial \bar{\phi}}{\partial z} = \frac{\partial \bar{\eta}}{\partial t}, \qquad \frac{\partial \bar{\phi}}{\partial t} + g\bar{\eta} = 0 \quad \text{on } z = 0.$$

In order for $\bar{\phi}$ to decay as $z \to -\infty$, we are forced to choose

$$\bar{\phi} = \bar{A}(k,t)e^{|k|z},$$

where $|k|\bar{A} = \partial\bar{\eta}/\partial t$, $\partial\bar{A}/\partial t = -g\bar{\eta}$, so that $\bar{A}$ satisfies

$$\frac{\partial^2 \bar{A}}{\partial t^2} = -g|k|\bar{A}.$$

Then, since $\bar{A} = 0$, and $\frac{\partial \bar{A}}{\partial t} = -g\bar{\eta}_0$ at $t = 0$, we find that

$$\bar{A} = -\bar{\eta}_0\sqrt{g/|k|}\,\sin\sqrt{g|k|}t \quad \text{and} \quad \bar{\eta} = \bar{\eta}_0\cos\sqrt{g|k|}t,$$

so that

$$\eta(x,t) = \frac{1}{2\pi}\int_{-\infty}^{\infty}\bar{\eta}_0(k)\cos\sqrt{g|k|}t\,e^{-ikx}\,dk. \qquad (4.18)$$

It is unfortunate that the integrand in this integral is, as it stands, a non-analytic function of k, but we recall from complex variable theory that we can define $|k| = \lim_{\varepsilon \to 0}\sqrt{k^2 + \varepsilon^2}$ as long as we define the "branch" of the function $\sqrt{k^2 + \varepsilon^2}$ correctly. This gives us a *theoretical* solution of the general gravity wave problem in two dimensions, but of what use is it? Only for a very few functions $\bar{\eta}_0$ will we be able to integrate (4.18) explicitly, and from the numerical point of view, (4.18) is a superposition of harmonic waves which is difficult to represent accurately with a computer, especially if $\bar{\eta}_0(k)$ is appreciable for large values of k. Nevertheless, there is an ingenious asymptotic method for determining how (4.18) behaves when x and t are large and this is the last piece of fundamental methodology to be described in this section.

It is well-known that integrals involving exponentials with large arguments can be evaluated asymptotically by *Laplace's method* and this theory is described by Hinch [10]. To illustrate the idea by an example, suppose we want to evaluate

$$I_1 = \int_{-\infty}^{\infty} A(k)e^{xg(k)}\,dk$$

asymptotically as $x \to +\infty$, where $g(k)$ is a real function that takes its largest value at $k = k_0$. Then, we find that[4]

$$I_1 \sim \frac{A(k_0)}{\sqrt{2\pi x|g''(k_0)|}}e^{xg(k_0)}\left(1 + O\left(\frac{1}{x}\right)\right).$$

However, the modification of this argument to consider an integral like (4.18) is more difficult. Nonetheless, if we consider

$$I_2 = \int_{-\infty}^{\infty} A(k)e^{ixg(k)}\,dk,$$

---

[4] Here, and subsequently, we use $\sim$ to denote expressions that are asymptotic to each other as defined by Hinch [10], for example; this is more precise than the cruder $\simeq$, which we use to mean "is approximately equal to."

where g is still real-valued, we can still assert that the main contribution to the integral as $x \to \infty$ usually comes from values of k for which $g'(k) = 0$. This results from the key observation that the real and imaginary parts of the integrand oscillate with a wavelength of $O(x^{-1})$ except near these critical values of k, where the wavelength is $O(x^{-1/2})$. The sketch in Figure 4.1 illus-

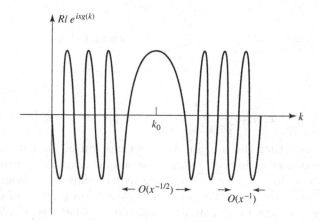

Fig. 4.1. $\mathrm{Rl}\,e^{ixg(k)}$ near a turning point of $g(k)$.

trates how, say, the real part of the integrand behaves. We can see that away from points where $g'(k) = 0$, the contribution to the integral from neighboring values of k will cancel out much more efficiently than it will in the neighborhood of such points. Thus, if $g(k)$ has one turning point at $k = k_0$, it can be shown that

$$I_2 \sim \frac{A(k_0)e^{ixg(k_0)}}{\sqrt{x}} \int_{-\infty}^{\infty} e^{ig''(k_0)s^2/2}\, ds.$$

Using contour integral methods, we can show that if $\lambda > 0$,

$$\int_{-\infty}^{\infty} e^{i\lambda s^2}\, ds = (1+i)\sqrt{\frac{\pi}{2\lambda}},$$

and, so,

$$\int_{-\infty}^{\infty} A(k)e^{ixg(k)}\, dk \sim \frac{\sqrt{\pi}A(k_0)e^{ixg(k_0)}(1+i)}{\sqrt{g''(k_0)x}} \qquad (4.19)$$

as $x \to \infty$, if $g''(k_0) > 0$. This method is called the *method of stationary phase* and the formula can easily be adapted to deal with large negative values of x or with cases where $g''(k_0) < 0$.

The estimate (4.19) immediately allows us to make predictions from (4.18). In order to see what happens for large x and t, we imagine an observer moving with constant speed V by writing $x = Vt$, and then we let $x, t \to \infty$, keeping

V constant. From (4.18), we are led to consider integrals of the form

$$\int_{-\infty}^{\infty} \bar{\eta}_0(k) e^{i(-kV \pm \sqrt{g|k|})t}\, dk.$$

The exponents $-kV \pm \sqrt{g|k|}$ have turning points at $k = \pm g/4V^2$ and we can see straightaway that, after a long time, the dominant waves seen by the observer have one or other of these wavenumbers. Using (4.19), we can see that, as $t \to \infty$,

$$\eta(Vt, t) \sim \mathrm{Rl}\left\{ \frac{C_1 e^{igt/4V} + C_2 e^{-igt/4V}}{\sqrt{t}} \right\}, \qquad (4.20)$$

for some complex constants C_i; the terms in C_1 and C_2 correspond to waves traveling in the positive and negative x directions, with $k = g/4V^2$ and $k = -g/4V^2$, respectively. Thus, as V increases, the frequency and wavenumber of the dominant observed waves decreases and their wavelength increases. Alternatively, a stationary observer far from the initial disturbance will see waves of gradually increasing frequency and wavenumber as time goes by. These predictions can be verified for the special case in which η_0 is localized[5] near $x = 0$ (see Exercise 4.4).

More generally, we notice that in the far field, waves of wavenumber k are only observed to dominate by an observer traveling at speed V if $|V| = \frac{1}{2}\sqrt{g/k}$; this speed is called the *group velocity* of waves of wavenumber k. This idea can be generalized to models where (4.18) is replaced by

$$\eta(x, t) = \frac{1}{2\pi} \int_{-\infty}^{\infty} \bar{\eta}_0(k) \cos \omega(k) t\, e^{-ikx}\, dk.$$

Then, the dominant contribution to the integral occurs when

$$V = \frac{d\omega}{dk}$$

and this is the mathematical definition of group velocity for such a wave model.

This is all dramatically different than the type of solution one gets when solving the acoustic wave equation (4.1) with localized initial data. In that case, $\omega(k) = \pm ck$ for any value of k, and the disturbance is eventually propagated at constant speed c without change of shape, so that results like (4.20) are not observed. The phenomenon we have just described for gravity waves is a manifestation of the *dispersive* nature of the system; dispersion occurs whenever the speed of waves varies with the wavenumber and we will consider this idea in more generality in the next section.

---

[5] They can also be observed by throwing a stone into a large pond.

4.4 *Dispersion and Group Velocity

We have already remarked that almost all of the models in Chapter 3 admit separable solutions that are products of exponential functions in space and time. Indeed, if we write the solution in each case as

$$\phi = \text{Rl}[Ae^{i\mathbf{k}\cdot\mathbf{x}-i\omega t}], \tag{4.21}$$

where, as usual, ω is the frequency and $\mathbf{k}$ is a "wavenumber vector" in the direction of the traveling wave, then we almost always find that ω satisfies a *dispersion relation* of the form

$$\omega = \omega(\mathbf{k}).$$

We have already seen explicit examples of such relations in Exercises 3.2 and 3.4–3.6, and in Section 4.3, we used the fact that for waves on deep water, $\omega = \sqrt{g|k|}$. In this case, we observed that the non-constancy of the group velocity $d\omega/dk$ led to the phenomenon of dispersion.

   We anticipate the fact that the stationary phase argument used in Section 4.3 can be applied to multiple Fourier integrals of the form

$$\iiint A(\mathbf{k})e^{i(\mathbf{k}\cdot\mathbf{x}-\omega t)}\, d\mathbf{k}$$

and will predict that the dominant contribution seen by an observer at $\mathbf{x} = \mathbf{V}t$ for large t comes from the values of $\mathbf{k}$ for which

$$\nabla_{\mathbf{k}}(\mathbf{k}\cdot\mathbf{V} - \omega(\mathbf{k})) = \mathbf{0}, \tag{4.22}$$

where $\nabla_{\mathbf{k}} = (\partial/\partial k_1, \partial/\partial k_2, \partial/\partial k_3)$ and $\mathbf{k} = (k_1, k_2, k_3)$. From (4.22), we see that

$$\mathbf{V} = \nabla_{\mathbf{k}}\omega \tag{4.23}$$

and this motivates us to define $\nabla_{\mathbf{k}}\omega$ as the group velocity of the waves with wavenumber $\mathbf{k}$. The *phase velocity* of these waves is

$$\frac{\omega\mathbf{k}}{|\mathbf{k}|^2}, \tag{4.24}$$

and (4.23) and (4.24) are only equal if $\omega = c|\mathbf{k}|$, where c is a constant. In all other circumstances, the wave speed varies with $\mathbf{k}$ and the system is dispersive.

   We now look at the models derived in Chapter 3 in the light of these ideas.

4.4.1 Dispersion Relations

Collecting together the dispersion relations for the systems considered in Chapter 3, we see that a number of these systems are indeed dispersive. With the exception of (ii) below, the following results emerge trivially from substituting solutions of the form (4.21) into the relevant field equations without imposing any boundary conditions; in case (ii), the boundary conditions are crucial.

(i) **Acoustic and Electromagnetic Waves.** Equations (3.6) and (3.32) lead to the dispersion relation

$$\omega^2 = c^2 |\mathbf{k}|^2,$$

so these waves are always non-dispersive.

(ii) **Surface Gravity Waves.** Two-dimensional waves satisfy (3.12) with boundary conditions (3.13) and (3.14) on water of depth h provided ϕ is proportional to $\mathrm{Rl}\cosh k_3(z+h)e^{i(k_1 x + k_2 y - \omega t)}$. This gives

$$\omega^2 = g|\mathbf{k}|\tanh|\mathbf{k}|h, \qquad (4.25)$$

where $|\mathbf{k}|^2 = k_1^2 + k_2^2$, and these waves are dispersive. The group velocity is given by

$$\mathbf{V} = \frac{g}{2\omega|\mathbf{k}|}(\tanh|\mathbf{k}|h + |\mathbf{k}|h\,\mathrm{sech}^2|\mathbf{k}|h)\mathbf{k}.$$

Note that as $h \to 0$, (4.25) reduces to

$$\omega^2 = gh|\mathbf{k}|^2,$$

so that waves on shallow water are non-dispersive and their phase speed is $\sqrt{gh}$.

(iii) **Inertial Waves on a Stratified Fluid.** A solution of type (4.21) only works for these waves if N is constant in (3.26). This can happen if ρ_0 is an exponential function of z, and in this case,

$$\omega^2 = \frac{N^2(k_1^2 + k_2^2)}{|\mathbf{k}|^2 - ik_3 N^2 g^{-1}}, \qquad (4.26)$$

which is clearly dispersive. Note that ω may now be complex when $\mathbf{k}$ is real, and we will discuss the implications of this at the end of the section.

(iv) **Rossby Waves.** For waves in a rotating fluid governed by (3.30), the dispersion relation is

$$\omega^2 = \frac{4\Omega^2 k_3^2}{|\mathbf{k}|^2}, \qquad (4.27)$$

and these waves are dispersive with group velocity

$$\mathbf{V} = \frac{2\Omega}{|\mathbf{k}|^3}(-k_1 k_3, -k_2 k_3, k_1^2 + k_2^2).$$

(v) **Elastic Waves.** Elastic waves are governed by (3.33), and if we write

$$\mathbf{u} = \mathrm{Rl}(\mathbf{A}e^{i\mathbf{k}\cdot\mathbf{x} - i\omega t}),$$

we find that the dispersion relation for longitudinal waves, where $\mathbf{A}$ is parallel to $\mathbf{k}$, is

$$\omega^2 = \left(\frac{\lambda + 2\mu}{\rho}\right)|\mathbf{k}|^2, \qquad (4.28)$$

whereas for transverse waves, for which $\mathbf{A}$ is perpendicular to $\mathbf{k}$,

$$\omega^2 = \frac{\mu}{\rho}|\mathbf{k}|^2. \tag{4.29}$$

Both of these waves are non-dispersive, but there is the possibility of "mode conversion" if energy is transferred from longitudinal waves to transverse waves or vice versa. This is an important phenomenon in seismic waves, as shown in Exercise 4.6.

Except for (ii), the above dispersion relations are all relevant for unconfined waves. However, the addition of boundaries can have a dramatic effect. In Section 4.2, we showed how the motion in a waveguide, where we solved the acoustic wave equation subject to boundary conditions on $y = 0$ and $y = b$, led to (4.14) and the dispersion relation

$$\omega^2 = c^2\left(k^2 + \frac{n^2\pi^2}{b^2}\right), \tag{4.30}$$

where n is a positive integer. Thus, waves in an acoustic waveguide are dispersive, and the group velocity is

$$V = \frac{ck}{\left(k^2 + \frac{n^2\pi^2}{b^2}\right)^{1/2}}.$$

Looking back at these examples, there are no obvious rules relating phase velocity and group velocity as defined by (4.23) and (4.24). They may be oriented in any direction relative to each other and either one may exceed the other in magnitude. However, there is one vitally important property of dispersive waves that can be discerned from the dispersion relation, namely whether ω is real or complex when k is real. Since all the above relations involve ω^2 only, there will be two roots $\pm\omega$ for any given $\mathbf{k}$ so that if the imaginary part of ω is non-zero, there will be solutions that grow exponentially in time. In fact, looking at the relations (4.25)–(4.30), we see that, fortunately, ω is real in all cases except for waves on a stratified fluid, when (4.26) shows that $\mathrm{Im}\,\omega$ is non-zero whenever k_3 is non-zero. When $\mathrm{Im}\,\omega$ is non-zero, the solution is *unstable* and this means that the linearized model from which the dispersion relation was derived will no longer be valid over sufficiently long timescales.

Another way of creating an instability arises if we change the sign of g in (4.25) so that the heavy fluid is above the lighter air. Then, ω is inevitably complex and this is the simplest example of the famous *Rayleigh–Taylor* instability. This is a particularly catastrophic instability because ω is imaginary for all k and is large for small wavelengths $2\pi/k$. As shown in Exercise 3.6, a very similar situation is revealed if we seek waves on the interface between two inviscid fluids moving parallel to each other and to the surface, but with different velocities. In this case, the instability is called the *Kelvin–Helmholtz* instability. Both of these instabilities can be stabilized by the introduction of surface tension at the interface; surface tension is a powerful mechanism at short wavelengths and allows the existence of capillary waves as described, for example, in Drazin and Reid [11] (see also Exercise 4.5).

4.4.2 Other Approaches to Group Velocity

Up to now, we have used large-time asymptotics as the motivation for introducing group velocity, but several other approaches are possible.

From the physical viewpoint, group velocity can be interpreted as the velocity with which the mean energy in a particular mode is transported; this idea is described in Lighthill [12]. A more elementary theoretical motivation comes from the following simple example, which gives an intuitive idea of the difference between phase and group velocity. We consider two sinusoidal wave trains of equal amplitude but slightly different wavenumber and frequency. The sum of these waves is

$$a \sin(k_1 x - \omega_1 t) + a \sin(k_2 x - \omega_2 t) = 2a \cos(\Delta k x - \Delta \omega t) \sin(kx - \omega t),$$

where $k_1 = k + \Delta k$, $k_2 = k - \Delta k$, $\omega_1 = \omega + \Delta \omega$, and $\omega_2 = \omega - \Delta \omega$. Thus, if Δk and $\Delta \omega$ are small, the result is a slowly modulated wave of amplitude $2a \cos(\Delta k x - \Delta \omega t)$, as shown in Figure 4.2. We can see that the so-called

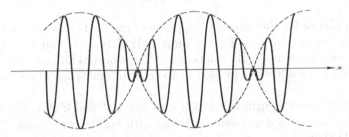

Fig. 4.2. Superposition of two harmonic waves.

"envelope" of the wave crests and troughs travels with speed $\Delta \omega / \Delta k$, which is approximately the group velocity of these waves.

Alternatively, we can take the asymptotic approach as in Section 4.3, but now we will use it directly on the equations rather than on the Fourier transform solution. This means that we need to use the *WKB expansion method* to determine the long-time or far-field solution of the linear wave model. To do this, we have to introduce a large artificial parameter ε^{-1} to represent the length and timescales and we rescale the independent variables by $x = \varepsilon^{-1} X$ and $t = \varepsilon^{-1} T$. Then, we write

$$\phi = \text{Rl}(\hat{A}(\mathbf{X}, T) e^{iu(\mathbf{X}, T)/\varepsilon}).$$

We note that we can do this only after the introduction of ε and taking the limit $\varepsilon \to 0$; otherwise, the functions u (the phase) and $\hat{A}$ (the amplitude) would not be uniquely defined[6] for a given ϕ.

Now, we can see that since ε is small,

$$\frac{\partial \phi}{\partial T} \sim \mathrm{Rl}\left[\left(i\frac{\hat{A}}{\varepsilon}\frac{\partial u}{\partial T}\right)e^{iu/\varepsilon}\right],$$

and

$$\frac{\partial^2 \phi}{\partial T^2} \sim \mathrm{Rl}\left[-\frac{\hat{A}}{\varepsilon^2}\left(\frac{\partial u}{\partial T}\right)^2 e^{iu/\varepsilon}\right],$$

to lowest order in ε. Now, from (4.21)

$$\frac{\partial \phi}{\partial t} = \mathrm{Rl}\left[-i\omega A e^{i\mathbf{k}\cdot\mathbf{x}-i\omega t}\right] \quad \text{and} \quad \frac{\partial^2 \phi}{\partial t^2} = \mathrm{Rl}\left[-\omega^2 A e^{i\mathbf{k}\cdot\mathbf{x}-i\omega t}\right],$$

and so we can identify $\partial u/\partial T$ with $-\omega$ and $\hat{A}$ with A in (4.21). Similarly, $\bar{\nabla}u$ can be identified with $\mathbf{k}$ in (4.21) to this approximation, where $\bar{\nabla} = (\partial/\partial X_1, \partial/\partial X_2, \partial/\partial X_3)$. Hence, if the dispersion relation is

$$\omega = \omega(\mathbf{k}),$$

we can immediately infer that u will satisfy

$$\frac{\partial u}{\partial T} + \omega(\bar{\nabla}u) = 0, \tag{4.31}$$

which is a first-order equation for the phase of the far-field solution. It is also possible to find an equation for the amplitude $\hat{A}$ (see Section 4.7). Even for non-dispersive systems, (4.31) will be a *nonlinear* partial differential equation; it only reduces to a linear equation in the case of one space dimension when $\omega = ck$.

Reverting to our identification of ω with $-\partial u/\partial T$ and $\mathbf{k}$ with $\bar{\nabla}u$, but now regarding ω and $\mathbf{k}$ as dependent variables with arguments $\mathbf{X}$ and T, we see from (4.31) that

$$-\bar{\nabla}\omega = \bar{\nabla}\frac{\partial u}{\partial T} = \frac{\partial}{\partial T}(\bar{\nabla}u) = \frac{\partial \mathbf{k}}{\partial T},$$

or[7]

$$\frac{\partial \mathbf{k}}{\partial T} + \sum_{i=1}^{3}\frac{\partial \omega}{\partial k_i}\bar{\nabla}k_i = \mathbf{0}. \tag{4.32}$$

---

[6] Note that with a conventional power series asymptotic expansion $\phi \sim \sum_0^\infty \varepsilon^n \phi_n$, we can define the terms recursively via $\phi_0 = \lim_{\varepsilon\to 0}\phi$, $\phi_1 = \lim_{\varepsilon\to 0}[(\phi - \phi_0)/\varepsilon]$, and so forth. Equally, for an expansion $\phi \sim e^{u/\varepsilon}\sum_{n=0}^\infty A_n\varepsilon^n$, where u is real, we can define $u = \lim_{\varepsilon\to 0}\varepsilon\log\phi$. However, in this case with u real, no such definition is possible because u now measures the oscillations in ϕ rather than its magnitude. The best that can be done is to say that $|\nabla u|^2 = \lim_{\varepsilon\to 0}(-\varepsilon^2\nabla^2\phi/\phi)$.

[7] Note that (4.32) can be thought of as "conservation of wavenumber" if ω is interpreted as "wavenumber flux."

We can most easily analyze this equation in one dimension when (4.32) reduces to

$$\frac{\partial k}{\partial T} + \frac{d\omega}{dk}\frac{\partial k}{\partial X} = 0,$$

which, since ω is only a function of k, has the general solution

$$k = F\left(X - \frac{d\omega}{dk}T\right)$$

for an arbitrary function F. Thus, k and ω are constant along the lines

$$X - \frac{d\omega}{dk}T = \text{constant},$$

and so, for large times, we again see that the waves with wavenumber k will dominate if we travel with the group velocity of the waves. We can also say that $\partial u/\partial T$ and $\bar{\nabla}u$ will be constant on these lines and so $u/\varepsilon = kx - \omega t$. All of this can be generalized to three dimensions and confirms, in a more general setting, the results in Section 4.3 for surface waves.

We will now put these general considerations on one side and consider some concrete situations in the frequency domain when boundaries are present. It turns out to be convenient to distinguish between problems that are stationary ($\omega = 0$) and those that are not, and we will start by considering the latter.

4.5 The Frequency Domain

4.5.1 Homogeneous Media

Thankfully, all of the phenomena to be discussed in this section can be illustrated with reference to the simplest *acoustic model* for monochromatic waves of prescribed frequency ω. Then, writing $\phi = \text{Rl}(\Phi(\mathbf{x})e^{-i\omega t})$ as usual, the equation for Φ is (4.13), which we write as

$$(\nabla^2 + k^2)\Phi = 0, \tag{4.33}$$

where $k = \omega/c$. This is *Helmholtz' equation*, which is an elliptic partial differential equation that appears to be a simple generalization of Laplace's equation. However, the mathematics associated with (4.33) turns out to be very different from the usual theories for Laplace's equation.

The first fundamental classification that must be made of waves in the frequency domain is the distinction between *interior* and *exterior* problems.

In the former case, we expect to have a boundary condition prescribed for Φ or $\partial\Phi/\partial n$, or some combination thereof, everywhere on a closed boundary and we are confronted with the problem of the solvability of (4.33) in D with, say, the boundary condition

$$\alpha\Phi + \beta\frac{\partial\Phi}{\partial n} = f \quad \text{on } \partial D.$$

If f, the forcing term, is identically zero, we have a classical eigenvalue problem for those values $k = k_i$ for which a non-trivial solution Φ_i exists. A great deal is known about such problems and much of it is described in Courant and Hilbert [13]. If, however, the forcing term f is non-zero, then we can use the eigenfunctions of the unforced problem to establish an integrability condition. This is another application of the Fredholm alternative mentioned in Section 4.2. Suppose, for example, that

$$\nabla^2 \Phi_i + k_i^2 \Phi_i = 0 \quad \text{in } D \qquad \text{and} \qquad \Phi_i = 0 \quad \text{on } \partial D$$

and

$$\nabla^2 \Phi + k^2 \Phi = 0 \quad \text{in } D \quad \text{with } \Phi = f \text{ on } \partial D.$$

From Green's second theorem, which states that

$$\int_D (\Phi \nabla^2 \Phi_i - \Phi_i \nabla^2 \Phi) dV = \int_{\partial D} \left(\Phi \frac{\partial \Phi_i}{\partial n} - \Phi_i \frac{\partial \Phi}{\partial n} \right) dS,$$

we see that if $k = k_i$, the solution Φ can only exist if

$$\int_{\partial D} f \frac{\partial \Phi_i}{\partial n} \, dS = 0.$$

If this condition is satisfied, then the solution for Φ will not be unique since it can only be determined to within a term $\lambda \Phi_i$, where λ is any constant. This is simply a more general way of stating the problem of resonance since $\omega_i = c k_i$ are the natural frequencies of the system.

Things are quite different when we examine the *exterior problem* in which boundary conditions are still imposed on ∂D but will not be sufficient by themselves to determine the physically relevant solution. Suppose, for example, we consider the problem of an oscillating sphere of radius a so that Φ is defined in $r > a$. The boundary condition

$$\phi = \text{Rl}(e^{-i\omega t}) \quad \text{on } r = a$$

implies that $\Phi = 1$ on $r = a$. Then, solving (4.33) with spherical symmetry leads to the solution

$$\Phi = \frac{A}{r} e^{ik(r-a)} + \frac{B}{r} e^{-ik(r-a)}, \tag{4.34}$$

where $A + B = a$. Even if we impose boundedness on Φ as $r \to \infty$, we are still short of a second equation relating A and B. What we have to remember is that by posing the problem in the frequency domain, we are considering a sphere that has been oscillating for a very long time so that the system has settled down to a solution periodic in time. However, we have implicitly assumed that while this motion was being set up, no other sources of waves have been sending disturbances *toward* the sphere from anywhere in $r > a$.

Thus, the waves described by (4.34) can only *radiate outward* from $r = a$ and this means that the second term in (4.34), which represents an *inward* traveling wave in the time domain, is not physically relevant.[8] Thus, the solution is given by taking $A = a$ and $B = 0$ and the argument which has led to this conclusion is referred to as "imposing a *radiation condition* at infinity."

This example suggests the hypothesis, which can be proved, that exterior frequency domain problems are well-posed (i.e., the solution exists, is unique, and depends continuously on the given data) as long as we impose the *Sommerfeld radiation condition*

$$r\left(\frac{\partial \Phi}{\partial r} - ik\Phi\right) \to 0 \qquad (4.35)$$

as $r \to \infty$. The premultiplier r is used in (4.35) because we know from (4.34) that $r\Phi$ is $O(1)$ as $r \to \infty$. The corresponding far-field behavior in one and two dimensions dictates the radiation conditions

$$\frac{\partial \Phi}{\partial x} - ik\Phi \to 0 \quad \text{as } x \to \infty$$

and

$$r^{1/2}\left(\frac{\partial \Phi}{\partial r} - ik\Phi\right) \to 0 \quad \text{as } r \to \infty,$$

respectively (where r is now a two-dimensional polar coordinate) and we will return to this dependence on dimensionality in Section 4.8 (see also Exercise 4.8).

This analysis of problems in which waves are radiating to infinity from a finite oscillator or "radiator" can, in principle, be extended to what are perhaps the most important frequency domain problems where incoming waves are *scattered* by a finite obstacle. Such problems arise naturally in applications ranging from oil exploration to radar and from harbor design to tomography.

4.5.2 Scattering Problems in Homogeneous Media

We suppose that a finite obstacle with boundary ∂D is "irradiated" or "insonified"[9] by a *plane wave* traveling in the positive x direction and given by $\Phi = Ae^{ikx}$. We take the boundary condition on ∂D to be $\Phi = 0$, which corresponds to a "hard" reflector, but we could equally well model a "soft" reflector by $\partial \Phi/\partial n = 0$. To solve this problem, all we need to do is consider $\hat{\Phi} = \Phi - Ae^{ikx}$ and solve Helmholtz' equation (4.33) for $\hat{\Phi}$ with $\hat{\Phi} = -Ae^{ikx}$ on ∂D and the appropriate Sommerfeld condition at infinity. This is another problem with a vast literature and we will mention briefly three aspects, each illustrated by a "canonical" problem.

[8] Note that had ϕ been written as $\mathrm{Rl}(\Phi e^{i\omega t})$, the first term would have been unacceptable.

[9] The term used depends on whether we are referring to light waves or sound waves.

(i) **Reflection.** Suppose that the body in the above problem is smooth and two dimensional. Locally, near any point on the body, the solution to (4.33) will consist of a combination of Fourier modes or plane waves so that

$$\Phi = \sum_{l,m} A_{lm} e^{ilx+imy},$$

where $l^2 + m^2 = k^2$ and we use l and m here rather than k_1 and k_2 to avoid "suffix clutter." Suppose we choose the tangent to the body at the point $(0,0)$ along the y axis, with the x axis along the inward normal. Then, it is clear that for every incoming wave of the form $A_{lm} e^{i(lx+my)}$, there must be a *reflected wave* of the form

$$-A_{lm} e^{i(-lx+my)}$$

if the boundary condition on $x = 0$ is to be satisfied.[10] Hence, every plane wave whose constant phase lines, or *wave fronts*, are $lx + my = $ constant induces a reflected plane wave whose constant phase lines are $lx - my = $ constant. For Helmholtz' equation, the normals to the wave fronts are called the *rays*; hence, we have *specular reflection* in which the angle of incidence of the incoming ray equals the angle of reflection of the outgoing ray, as shown in Figure 4.3.

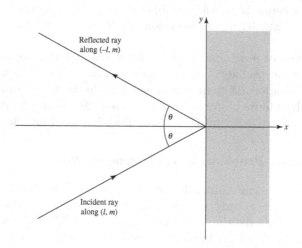

Fig. 4.3. Wave reflection at a plane boundary.

(ii) **Refraction.** The phenomenon of refraction occurs when waves in a medium in which the sound speed is c_0 irradiate a body that is capable of transmitting waves through its interior, where the ambient

---

[10] The amplitude of the reflected wave would be different if the boundary condition was of the form $\alpha \Phi + \beta(\partial \Phi / \partial n) = 0$.

speed of sound is c_1. Again, thinking of the wave $A_{lm}e^{i(lx+my)}$ impinging on the surface $x = 0$, we will now have a reflected wave given by
$Be^{i(-lx+my)}$ in $x < 0$ and a transmitted wave $Ce^{i(l_1x+m_1y)}$ in $x > 0$.
Here, $l_1^2 + m_1^2 = \omega^2/c_1^2 = (c_0^2/c_1^2)k^2$, and no matter what other continuity
conditions we impose on $x = 0$, we must have a "continuity of wavenum-
ber" so that $m_1 = m$. If θ_r is the angle of refraction and θ_i is the angle of
incidence, we see from Figure 4.4 that

$$\tan\theta_i = \frac{m}{l} = \frac{m}{\sqrt{k^2 - m^2}}, \qquad \sin\theta_i = \frac{m}{k},$$

and

$$\tan\theta_r = \frac{m}{l_1} = \frac{m}{\sqrt{(c_0^2/c_1^2)k^2 - m^2}}, \qquad \sin\theta_r = \frac{c_1 m}{c_0 k},$$

and so

$$\sin\theta_r = \frac{c_1}{c_0}\sin\theta_i, \tag{4.36}$$

which is *Snell's law of refraction*. This condition leads to the possibility
of *total internal reflection* if $(c_1/c_0)\sin\theta_i > 1$, in which case the solution decays exponentially on $x > 0$ (see Exercise 4.9). The amplitudes of
the reflected and refracted waves will depend on the exact form of the
boundary conditions that are applied.

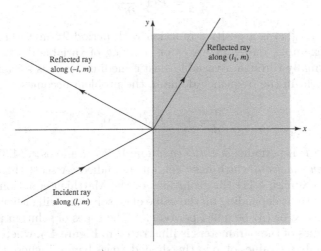

Fig. 4.4. Refraction at an interface.

The phenomena of reflection and refraction apply locally to all scatterers
that are smooth enough to have a tangent plane to which the irradiation
is not tantential. Hence, they give good intuition as to the way in which
many scatterers respond to incoming waves. Alas, non-smooth scatterers
often radiate most intensely, which is explained by the next phenomenon.

(iii) **Diffraction**. Diffraction is a much more complicated phenomenon that cannot be described simply in terms of a few plane waves, as was possible for reflection and refraction. It occurs when a plane wave is incident on an obstacle in a situation where *infinitely many* new plane waves are generated. This can happen, for instance, when the obstacle is not smooth or when the incident wave is tangential to the body.

It is possible, by using ingenious asymptotics, to solve both of these problems and to write down the intensity of the diffracted field at infinity, and we have indicated schematically the directions of the scattered field on the plots in Figures 4.13 and 4.14. In each case, a radiation condition has to be applied to the solution after the incident field has been subtracted out. A better idea of the scattering that can be expected will emerge from the discussion in Section 4.7 on high-frequency waves.

4.5.3 Inhomogeneous Media

One especially interesting phenomenon is the propagation of waves through materials whose properties vary smoothly. A one-dimensional analysis of the propagation of waves through a periodic medium reveals some of the effects that can occur. We consider a conceptual generalization of (3.4) in the form

$$\frac{\partial^2 \phi}{\partial x^2} = P(x)\frac{\partial^2 \phi}{\partial t^2},$$

where $P = 1/c_0^2(x)$ is a positive function with period 2π in x. This equation could, for example, represent waves on a string of variable density or waves passing normally through a variable elastic medium such as seismic waves in stratified rock. In the frequency domain, the problem becomes

$$\frac{d^2 \Phi}{dx^2} + \omega^2 P(x)\Phi = 0, \tag{4.37}$$

which, when P is periodic, is *Hill's equation*. If $P = a + b\cos x$, (4.37) becomes the *Matthieu equation*, which has been much studied (Arscott [14]). It can be shown (see Exercise 4.11) that solutions of the Matthieu equation are rarely periodic and that, depending on the value of ω, solutions will either (i) grow or decay as $x \to \infty$ or (ii) be quasi-periodic.[11] The types of solution possible for different values of the parameters is illustrated in Figure 4.5, where waves can propagate only for values of ω in the shaded "pass bands," where the solution is quasi-periodic. The key feature is that, in the so-called "stop bands," which are the complements of the pass bands in Figure 4.5, the medium causes waves to decay exponentially if they try to propagate in the x direction.

It can sometimes be instructive to model a one-dimensional smoothly varying inhomogeneous medium as a composition of parallel layers and use the

---

[11] A quasi-periodic solution consists of a sum of periodic terms with non-commensurate periods.

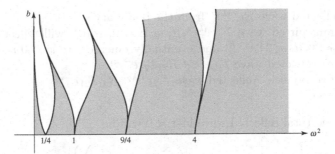

Fig. 4.5. Solutions of the Mathieu equation for small b.

ideas of Section 4.5.2 at the interface between each layer. We will just look at one layer, the analysis of which is important in the transmission and reflection of waves through thin coatings. Here, we only consider acoustic waves propagating in the x direction through a medium of density ρ_1 and consider the effect of introducing a layer of density ρ_2 in $0 < x < h$. Concerning the transition from medium 1 to medium 2 at $x = 0$, if the potentials in the frequency domain are Φ_1 in $x < 0$ and Φ_2 in $x > 0$, the conditions to be satisfied at $x = 0$ are

$$\rho_1 \Phi_1 = \rho_2 \Phi_2 \quad \text{and} \quad \frac{d\Phi_1}{dx} = \frac{d\Phi_2}{dx},$$

which represent the continuity of pressure and velocity at the interface. If the incident wave is given by

$$\Phi_1 = e^{ik_1 x},$$

we could proceed directly by solving Helmholtz' equation in the three regions $x < 0, 0 < x < h$, and $h < x$, with appropriate boundary conditions, as in Exercise 3.10. However, here we use an alternative approach in which the waves are almost thought of as "particles." To do this, we note that the incident wave will engender a reflected wave $R_{12} e^{-ik_1 x}$ in $x < 0$ and a transmitted wave $T_{12} e^{ik_2 x}$ in $h > x > 0$, where

$$R_{12} = \frac{Z_2 - Z_1}{Z_2 + Z_1}, \qquad T_{12} = \frac{\rho_1}{\rho_2} \left(\frac{2Z_2}{Z_1 + Z_2} \right)$$

and where $Z_i = \rho_i / k_i$ is called the *wave impedance*. Similarly, the reflection and transmission coefficients for a wave going from medium 2 to medium 1 are

$$R_{21} = \frac{Z_1 - Z_2}{Z_1 + Z_2} = -R_{12} = -R,$$

say, and

$$T_{21} = \frac{\rho_2}{\rho_1} \left(\frac{2Z_1}{Z_1 + Z_2} \right) = \frac{\rho_2}{\rho_1} (1 - R).$$

Now, if the incident wave $\Phi_1 = e^{ik_1 x}$ impinges on the layer, it will suffer repeated reflections and transmissions as follows:

(i) The reflected wave $Re^{-ik_1 x}$ from the boundary $x = 0$.

(ii) The transmitted wave $T_{12}e^{ik_2 x}$ from $x = 0$, which will reflect at $x = h$ as $T_{12}e^{ik_2 h}R_{21}e^{-ik_2(x-h)}$ and eventually emerge into $x < 0$ as the "one-bounce" reflected wave $T_{12}e^{2ik_2 h}R_{21}T_{21}e^{-ik_1 x}$.

(iii) The "two-bounce" reflected wave $T_{12}e^{4ik_2 h}R_{21}^3 T_{21}e^{-ik_1 x}$ and so on.

Thus, the total reflected wave in $x < 0$ will be

$$\left\{ R - T_{12}T_{21}Re^{2ik_2 h}\left(\sum_{n=0}^{\infty} R^{2n}e^{2ink_2 h} \right) \right\} e^{-ik_1 x},$$

which can be simplified to

$$\frac{R(1 - e^{2ik_2 h})}{(1 - R^2 e^{2ik_2 h})} e^{-ik_1 x}.$$

Thus, we can see that the layer does not reflect at all if $k_2 h = n\pi$, where n is an integer; when $k_2 h = \pi$, this is called a "half-wavelength" non-reflecting layer.

4.6 Stationary Waves

There are two frequency domain situations that are much easier to analyze than the general cases considered in the previous section. These occur when the wavelength $2\pi/k = 2\pi c/\omega$ is either very large or very small compared to a typical length scale of interest, and in either case, we can use asymptotic methods to analyze a number of problems. The first case occurs, for example, in acoustics, where wavelengths may be a few meters, and the second case arises frequently in electromagnetism, especially in optics, where the wavelengths are about 10^{-9} m. The case $\omega = 0$ corresponds to stationary waves and we start by considering a small k since this means that Helmholtz' equation (4.33) is a *regular perturbation*[12] of Laplace's equation. However, this limit does not usually reveal any very interesting phenomena[13] unless there is a large source of energy that can be tapped, as is the case when the wave-bearing medium is moving bodily, and all the following problems fall into this category.

Hence, we will now revisit some examples that were introduced in Chapter 3 and consider surface gravity waves on a uniform stream and acoustic waves in a pipe of slowly varying cross-section and in a medium flowing past a thin obstacle. The consideration of problems in which k is large will be left to Section 4.7.

---

[12] For a definition of a regular perturbation, see Hinch [10].

[13] See Exercise 4.12 on the Helmholtz resonator for one interesting case.

4.6.1 Stationary Surface Waves on a Running Stream

We have seen in Exercise 3.4 that the dispersion relation for two-dimensional gravity waves on the surface of a stream of depth h moving with constant speed U is

$$(\omega \pm Uk)^2 = gk \tanh kh,$$

and it was also shown that stationary waves with $\omega = 0$ can occur only if $U^2 < gh$, in which case the flow is said to be *subcritical*. Waves of this type can often be observed upstream of an obstruction in a river and such waves will only be independent of time in a fairly sluggish flow.

The parameter $U/\sqrt{gh}$ is important in many free surface flows and it is called the *Froude number*,[14] F. Flows with $F > 1$ are called *supercritical*.

The solution is even more interesting in three dimensions when stationary waves with wavenumber k_1 in the $\mathbf{U}$ direction and wavenumber k_2 in a perpendicular direction satisfy the relation

$$U^2 k_1^2 = g|\mathbf{k}| \tanh |\mathbf{k}| h, \tag{4.38}$$

where $|\mathbf{k}| = (k_1^2 + k_2^2)^{1/2}$ (Exercise 3.5). Now, even in the simplest case of infinitely deep water, we can find infinitely many real values for $\mathbf{k}$ given U and g. When $h \to 0$, the equation reduces to

$$(U^2 - gh)k_1^2 = ghk_2^2,$$

and, as long as $k_2 \neq 0$, there will now be real solutions for $\mathbf{k}$ only in supercritical flow, when $U^2 > gh$. We will encounter this dispersion relation again in Section 4.6.3 when we study stationary waves in supersonic flow.

The beautiful pattern of waves behind a ship can be analyzed using the dispersion relation (4.38). If a ship is traveling with speed U on deep water, then the general solution for the wave elevation is

$$\eta = \int_{-\infty}^{\infty} F(k_1) e^{-i(k_1 x + k_2 y)} \, dk_1, \tag{4.39}$$

where x and y are measured from the ship, along and perpendicular to the direction of travel and, from (4.38) with $h = \infty$, k_2 is related to k_1 by

$$U^2 k_1^2 = g(k_1^2 + k_2^2)^{1/2}; \tag{4.40}$$

hence, the integral in (4.39) is taken over values of k_1 satisfying $|k_1| > g/U^2$. The function F is a function that will depend on the flow in the vicinity of the ship, but its precise form is unimportant when we look at waves far from the ship. This is because we can again use the method of stationary phase (Section 4.3) to estimate the form of η for large values of x and y. On writing $y = \lambda x$ and letting $x \to \infty$ while keeping λ constant, we can apply (4.19) to

---

[14] This is pronounced "Frowd."

see that the dominant contribution to the integral in (4.39) will come from values of k_1 for which

$$\frac{d}{dk_1}(k_1 + \lambda k_2) = 0. \tag{4.41}$$

From (4.40),

$$k_2 = \pm k_1 \left(\frac{U^4 k_1^2}{g^2} - 1 \right)^{1/2},$$

and so (4.41) leads to

$$\lambda = \mp \left(\frac{U^4 k_1^2}{g^2} - 1 \right)^{1/2} \left(\frac{2U^4 k_1^2}{g^2} - 1 \right)^{-1}. \tag{4.42}$$

This gives real values for k_1 only if $|\lambda| < 1/2\sqrt{2}$ and so we can see immediately that the waves are all contained in a wedge behind the ship of angle $\tan^{-1}\left(1/2\sqrt{2}\right) = \sin^{-1}(\frac{1}{3})$. The angle of this wedge is thus independent of the speed or any other properties of the ship. The pattern of the wave crests, sketched in Figure 4.6, can also be calculated. All we need to do is to plot the level curves of the phase $k_1 x + k_2 y$ remembering that k_1 and k_2 are related via (4.40) and that k_1 is a function of $\lambda = y/x$ from (4.42). The details are

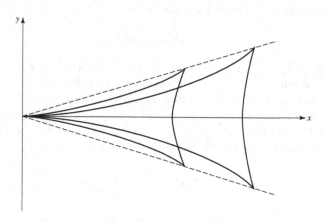

Fig. 4.6. Wavecrest pattern for ship waves.

left to Exercise 4.14. The edge of the wedge shown in Figure 4.6 is a kind of "envelope" of these wave crests or wave fronts, and we will see another example of this phenomenon in Section 4.7.

4.6.2 Steady Flow in Slender Nozzles

We next consider the homentropic, irrotational, steady flow of a gas along a pipe or nozzle which is aligned with the x axis and is such that the cross-sectional area of the pipe $A(x)$ varies slowly with x. We could proceed by studying the steady form of (2.6)–(2.8) and linearizing about a unidirectional flow (Exercise 4.15), but it is easier, if less systematic, to revert to a "control volume"[15] approach for mass conservation. Assuming that the flow is unidirectional to a first approximation, we use conservation of mass on the volume shown in Figure 4.7 to say that

$$\rho u A = \text{constant.} \tag{4.43}$$

Also, by Bernoulli's equation (2.25),

$$\frac{1}{2}u^2 + \frac{\gamma p}{(\gamma - 1)\rho} = \text{constant,} \tag{4.44}$$

where we have used the result

$$\frac{p}{p_0} = \left(\frac{\rho}{\rho_0}\right)^{\gamma} \tag{4.45}$$

since the flow is homentropic. Moreover, since the flow is irrotational, u is approximately a function of x alone and, hence, so are ρ and p. If we now

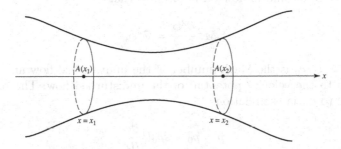

Fig. 4.7. Control volume for flow in a nozzle.

linearize this system by writing $A = A_0 + \bar{A}$, $\rho = \rho_0 + \bar{\rho}$, $u = u_0 + \bar{u}$, and $p = p_0 + \bar{p}$ where A_0, ρ_0, u_0, and p_0 are all constants and $\bar{A}$, $\bar{\rho}$, $\bar{u}$, and $\bar{p}$ are small perturbations, we get

$$\frac{\bar{\rho}}{\rho_0} + \frac{\bar{A}}{A_0} + \frac{\bar{u}}{u_0} = 0$$

and

$$u_0\bar{u} + \frac{\gamma}{\gamma - 1}\frac{\bar{p}}{\rho_0} - \frac{c_0^2\bar{\rho}}{(\gamma - 1)\rho_0} = 0.$$

---

[15] This is a very useful engineering technique in which the fluid is divided into regions, possibly large ones, and estimates are made for the *global* changes in mass, momentum and energy in these regions in terms of their boundary values.

Also, from (4.45), $\bar{p} = c_0^2 \bar{\rho}$, where $c_0^2 = \gamma p_0 / \rho_0$, and so, solving for $\bar{\rho}$ and $\bar{u}$ gives

$$\frac{\bar{\rho}}{\rho_0} = \frac{M^2 \bar{A}}{A_0(1 - M^2)}$$

and

$$\frac{\bar{u}}{u_0} = \frac{\bar{A}}{(M^2 - 1)A_0},$$

where $M = u_0 / c_0$ is the Mach number of the basic flow. These formulas show that our intuition that u increases (decreases) when A decreases (increases), which is true when $M < 1$, must be reversed when $M > 1$. Even worse, they show that the linear theory will be invalid when M is close to unity. Thus, this linearized theory needs careful reconsideration and we will return to discuss the nonlinear system (4.43)–(4.45) for ρ, p, and u in more detail in Section 6.2.3 of Chapter 6.

4.6.3 Compressible Flow past Thin Wings

It was shown in Chapter 3 that the steady linearized equation for a small disturbance to a uniform flow $U\mathbf{i}$ is (3.7), so that

$$M^2 \frac{\partial^2 \phi}{\partial x^2} = \nabla^2 \phi, \tag{4.46}$$

where $M = U/c_0$ is the Mach number of the undisturbed flow and $\varepsilon \phi$ is the correction to the velocity potential of the undisturbed flow. The pressure is connected to ϕ via the relation

$$p = p_0 - \varepsilon \rho_0 U \frac{\partial \phi}{\partial x}, \tag{4.47}$$

where the small parameter ε characterizes the size of the disturbance to the uniform flow. These equations can now be applied to the flow past a thin two-dimensional wing which is nearly aligned with the flow. We suppose the upper and lower surfaces of the wing are given by

$$y = \varepsilon f_\pm(x) \quad \text{for } 0 < x < l,$$

so that the boundary condition on the wing is

$$\varepsilon f'_\pm(x) = \frac{\varepsilon \partial \phi / \partial y}{U + \varepsilon \partial \phi / \partial x} \quad \text{on } y = \varepsilon f_\pm(x),$$

and the linear approximation reduces this to

$$\frac{\partial \phi}{\partial y} = U f'_\pm(x) \quad \text{on } y = 0 \pm \text{ for } 0 < x < l. \tag{4.48}$$

In order for the linearized model to be valid, we need to assume $f'_\pm(x)$ is of $O(1)$.

As can be seen immediately from equation (4.46), the solution to this problem depends crucially on whether M is greater or less than 1.

(i) **Subsonic Flow, $M < 1$.** When M is less than 1, (4.46) is elliptic and, writing $\beta^2 = 1 - M^2$, we have to solve

$$\beta^2 \frac{\partial^2 \phi}{\partial x^2} + \frac{\partial^2 \phi}{\partial y^2} = 0 \qquad (4.49)$$

subject to the boundary conditions (4.48) and the condition that $|\nabla \phi| \to 0$ at infinity. By rescaling $y = Y/\beta$, (4.49) becomes Laplace's equation, and if we also write $\phi = (1/\beta)\Phi(x, Y)$, then (4.48) remains

$$\frac{\partial \Phi}{\partial Y} = U f'_\pm(x) \quad \text{on } Y = 0\pm.$$

Thus, we see that Φ is the potential for an *incompressible* flow past the same thin wing.

It is easy to see that since $\log(x^2 + \beta^2 y^2)$ is a solution of (4.49), a possible form for the solution of this problem is

$$\phi = \int_0^l g(\xi) \log((x - \xi)^2 + \beta^2 y^2) \, d\xi \qquad (4.50)$$

for some function g, which is equivalent to a "source distribution" along the x axis. Morever, when we let $y \to 0+$ for $0 < x < l$, we find (Exercise 4.16) that

$$\frac{\partial \phi}{\partial y} \to 2\beta\pi g(x),$$

and so the choice

$$g(x) = \frac{U}{2\pi\beta} f'_+(x) \qquad (4.51)$$

will make (4.50) satisfy the boundary condition on the top of the wing.

It is only if the wing is symmetric, however, that the solution of the form (4.50) with g given by (4.51) will also satisfy the boundary condition as $y \to 0-$ and thus provide the solution we seek. For an unsymmetrical wing, we need to introduce a distribution of vortices as well as sources along the x axis (Exercise 4.16).

This idea of using distributions of sources and vortices to represent a thin two-dimensional wing is exactly the same as that used in incompressible theory and, indeed, we can apply all of the theory of incompressible flow past thin bodies to this problem. In particular, D'Alembert's paradox implies that there will be no forces on the wing if there is no circulation. Hence, for a symmetric wing, there will be neither drag nor lift on the wing in subsonic flow. However, for an asymmetric wing with the Kutta–Joukowski condition applied at the trailing edge, there will be a lift and we can relate the force in the compressible case to that in the incompressible case. In the linearized approximation, the lift L_c in the compressible case will be

$$L_c = \int_0^l (-p_+ + p_-)\, dx,$$

where $p_\pm$ are the pressures just above and below the wing. Using (4.47), we have

$$L_c = \varepsilon \rho_0 U \int_0^l \left(\frac{\partial \phi_+}{\partial x} - \frac{\partial \phi_-}{\partial x} \right) dx = \frac{1}{\beta} L_i, \qquad (4.52)$$

where L_i is the lift in the incompressible case. For incompressible flow past a flat plate of length l at a small angle $-\varepsilon$ to the flow, it can be shown (Acheson [5]) that the circulation is $\pi U l \varepsilon$, and, hence, by the Kutta–Joukowski theorem, the lift on the wing in compressible flow will be $\rho_0 \pi U^2 l \varepsilon / \beta$, and there will be no drag.[16] This result is yet another manifestation of the breakdown of linear theory when $M \to 1$ and $\beta \to 0$.

(ii) **Supersonic Flow, $M > 1$.** When M is greater than 1, (4.46) is hyperbolic and we can write down the general solution of this wave equation as

$$\phi = F(x - By) + G(x + By),$$

where $B^2 = M^2 - 1$. However, because the equation is hyperbolic, we need to impose different boundary conditions compared to the subsonic case. On physical grounds, we assert that there will be no *upstream influence* due to the wing and, therefore, we impose Cauchy data $\partial \phi / \partial x = \phi = 0$ on $x = 0$. From this, it follows that

$$\phi = \phi_+ = F(x - By) \quad \text{in } y > 0$$

and

$$\phi = \phi_- = G(x + By) \quad \text{in } y < 0.$$

Now, applying the boundary conditions (4.48) gives

$$-BF'(x) = U f'_+(x), \qquad BG'(x) = U f'_-(x)$$

for $0 < x < l$, and the solution is

$$\phi_+ = -\frac{U}{B} f_+(x - By) \quad \text{for } 0 < x - By < l, \ y > 0 \qquad (4.53)$$

and

$$\phi_- = \frac{U}{B} f_-(x + By) \quad \text{for } 0 < x + By < l, \ y < 0. \qquad (4.54)$$

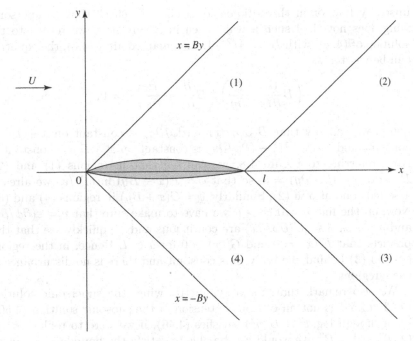

Fig. 4.8. Supersonic flow past a thin wing.

These formulas suggest that there will be zones of silence both in front of and behind the wing, as shown in Figure 4.8, and we will return to this point shortly.

First, we use formula (4.52) to find the lift explicitly as

$$L = \frac{\rho_0 U \varepsilon}{B} \int_0^l (-f'_+(x) - f'_-(x))\, dx$$

$$= -\frac{\rho_0 U^2 \varepsilon}{B}(f_+(l) + f_-(l)),$$

assuming that $f_+(0) = f_-(0) = 0$. Thus, for a flat plate at incidence where $f_+(x) = f_-(x) = -x$, the lift will be non-zero and its value is $2\rho_0 U^2 \varepsilon l/B$, even without the assumption of a "trailing edge" condition. This is in marked contrast to the subsonic case, as is the fact that the drag on a flat plate now comes out to be $2\rho_0 U^2 \varepsilon^2 l/B$ (Exercise 4.19). The fact that the forces on the wing increase dramatically as $B \to 0$ or as $\beta \to 0$ in (4.52) caused one of the major difficulties in the early days of supersonic flight.

It is well known that in incompressible flow, thin lifting wings leave wakes of concentrated vorticity behind them either in two-dimensional

---

[16] More generally, the lift on an arbitrary thin wing is $\rho_0 U \Gamma/\beta$, where Γ is the circulation around the wing in the incompressible case.

unsteady flow or in three-dimensional flow. To check that a supersonic wing does not shed such a wake even in steady flow, we return to the solution of (4.46) with $B^2 = M^2 - 1$. As remarked after (4.3), the equation can be written as

$$\left(B\frac{\partial}{\partial x} - \frac{\partial}{\partial y}\right)\left(B\frac{\partial}{\partial x} + \frac{\partial}{\partial y}\right)\phi = 0,$$

hence, we can say that $B(\partial\phi/\partial x) + (\partial\phi/\partial y)$ is constant on $x + By = $ constant and $B(\partial\phi/\partial x) - (\partial\phi/\partial y)$ is constant on $x - By = $ constant.[17] Now, referring to Figure 4.8, we can see that in regions (1) and (2), $B(\partial\phi/\partial x) + (\partial\phi/\partial y) = 0$, so that $\phi = F(x - By)$ in (1) (as we already asserted) and also in (2). Similarly, $\phi = G(x + By)$ in regions (4) and (3). Now, on the line $y = 0$, $x > l$, we have to make sure that $v = \varepsilon(\partial\phi/\partial y)$ and $p = -p_0 - \varepsilon\rho_0 U(\partial\phi/\partial x)$ are continuous and we quickly see that this predicts that $F'(x) = 0$ and $G'(x) = 0$ for $x > l$. Hence, in the regions (2) and (3) behind the body, ϕ is constant and there is no disturbance in these regions.

We also remark that, for a symmetric wing, the supersonic solution (4.53)–(4.54) is not an obvious extension of the subsonic solution (4.50). Even though $\log(x^2 - B^2 y^2)$ satisfies (4.46), if we were to replace β^2 in (4.50) with $-B^2$, we would not be able to satisfy the boundary condition on $y = 0$. However, we can write (4.53)–(4.54) as

$$\phi_\pm(x, y) = \mp\frac{U}{B}\int_0^l f'_\pm(\xi)H(x - \xi \mp By)\,d\xi, \qquad (4.55)$$

where H is the Heaviside function defined by

$$H(x) = \begin{cases} 0, & x < 0 \\ 1, & x > 0. \end{cases}$$

Note that the upper limit of the integral in (4.55) is $x \mp By$ if $0 < x \mp By < l$, and there is no velocity perturbation downstream of the characteristics through the trailing edge.

This question of downstream influence is much more interesting, even for non-lifting bodies, in three dimensions and we will consider the linearized flow past a slender axisymmetric body such as a rocket in Section 4.6.4. Before doing so, however, we must make two caveats about our linearized two-dimensional model. The first is that our predictions of infinite forces on aerofoils as $M \to 1$ in both subsonic and supersonic flow means that the linear model is invalid when $M^2 - 1$ is small. This fact is not surprising because, in this limit, the term $(M^2 - 1)(\partial^2\phi/\partial x^2)$ in (4.46) becomes so small as to be comparable with the nonlinear terms that we have neglected. Quite a complicated asymptotic procedure is needed to

[17] $B(\partial\phi/\partial x) \pm (\partial\phi/\partial y)$ are called *Riemann invariants* (see Ockendon et al. [9]).

derive a consistent limit in this case, and because the upshot is a nonlinear model, we will defer its derivation to Section 6.3.1 of Chapter 6.

A similar caution applies when M is large, even though the slope of the body is of $O(\varepsilon)$, where ε is small. When $M\varepsilon$ is of $O(1)$, the region of influence of the body, which is bounded by $x = \pm By$ in two dimensions, becomes very thin. However, this inevitably introduces nonlinearity into the problem again and so we leave further discussion to Section 6.3.4 of Chapter 6.

4.6.4 Compressible Flow past Slender Bodies

In this section, we consider the flow of a uniform stream past a slender axisymmetric body of length l given by $r = \varepsilon R(x)$ in cylindrical polars (r, θ, x). We will begin by considering a free stream which is aligned with the axis of the body and leave the case of a body at a small angle of incidence to the free stream to Exercise 4.18.

The equation to be solved is still (4.46), which can be written in cylindrical polar coordinates (r, θ, x) as

$$(M^2 - 1)\frac{\partial^2 \phi}{\partial x^2} = \frac{\partial^2 \phi}{\partial r^2} + \frac{1}{r}\frac{\partial \phi}{\partial r} + \frac{1}{r^2}\frac{\partial^2 \phi}{\partial \theta^2}. \tag{4.56}$$

Since the velocity of the flow is $(0, 0, U) + \varepsilon \nabla \phi$ and the normal to the body is $(1, 0, -\varepsilon R'(x))$, the linearized form of the boundary condition on the body is

$$\frac{\partial \phi}{\partial r} = U R'(x) \tag{4.57}$$

on $r = \varepsilon R(x)$. In two dimensions, we obtained good results by applying this condition on the x axis to give (4.48) and this certainly made the mathematics simpler. Now, we will not be so lucky.

The need for modification becomes apparent when we regard the solution (4.50) as a distribution along $y = 0$, $0 < x < l$, of simple source solutions of (4.49) of the form $\log(x^2 + \beta^2 y^2)$. As shown in Exercise 4.16(i), this distribution leads to the integral in (4.50) having a *power series expansion* in y for both $y > 0$ and $y < 0$, and this is what allows the boundary condition (4.48) to be applied on $y = 0$. When we try to distribute *axisymmetric* source solutions of (4.56), of the form $(x^2 + \beta^2 r^2)^{-1/2}$, along the line $r = 0$, $0 < x < l$, we find that $\partial \phi / \partial r$ inevitably approaches infinity as $r \to 0$. This is because the integral of $\partial \phi / \partial r$ around any small circle enclosing the body has to stay finite as the radius of the circle tends to zero. Thus, we will have to be careful when we apply the boundary condition (4.57).

(i) **Subsonic Flow, $M < 1$.** Having already observed that if $\beta^2 = 1 - M^2$, there is an elementary axisymmetric solution to (4.56) in the form $(x^2 + \beta^2 r^2)^{-1/2}$, we try a source distribution of the form

$$\phi = \int_0^l \frac{h(\xi)\, d\xi}{((x - \xi)^2 + \beta^2 r^2)^{1/2}}. \tag{4.58}$$

Now, it is easy to show (Exercise 4.16(i)) that as $r \to 0$,

$$\frac{\partial \phi}{\partial r} \sim -\frac{2h(x)}{r},$$

and so, applying the boundary condition (4.57) on $r = \varepsilon R$, we find that

$$h(x) = -\frac{\varepsilon}{2} U R(x) R'(x). \tag{4.59}$$

Thus, we have obtained the solution to axisymmetric slender body theory in which the disturbance to the flow variables turns out to be of $O(\varepsilon^2)$ compared to the undisturbed quantities in spite of the fact that the body width is of $O(\varepsilon)$ compared to its length.

(ii) **Supersonic Flow, $M > 1$.** We recall that in going from subsonic flow to supersonic flow the two dimensions, we could not simply replace β^2 in (4.50) by $-B^2$. However, the fact that the boundary conditions in slender body flow have to be applied on $r = \varepsilon R$ rather than on $r = 0$ means that we may now be able to generalize the idea leading to (4.59) in order to solve the supersonic problem. Since the function

$$\psi = \begin{cases} 0, & |x| < Br \\ (x^2 - B^2 r^2)^{-1/2}, & |x| > Br \end{cases} \tag{4.60}$$

formally satisfies (4.56), except when $|x| = Br$, we try

$$\phi = \int_0^{x-Br} \frac{m(\xi)\, d\xi}{[(x - \xi)^2 - B^2 r^2]^{1/2}} \tag{4.61}$$

for $x > Br$. The choice of the upper limit is not only suggested by the form of solution (4.60) but is also confirmed by Exercise 4.25 and by the physical expectation that the supersonic solution at $P(r, \theta, x)$ will only depend on the body shape upstream of the point A, where the length OA is $x - Br$, as shown in Figure 4.9.

Now, letting $r \to 0$ in (4.61), we find that $\phi \sim m(x) \cosh^{-1}(x/Br)$, which means, following Exercise 4.16(i), that $\phi \sim m(x) \log r$ to lowest order; thus, using (4.57),

$$m(x) = \varepsilon U R(x) R'(x). \tag{4.62}$$

Hence, the solution is

$$\phi = \int_0^{x-Br} \frac{\varepsilon U R(\xi) R'(\xi)\, d\xi}{((x - \xi)^2 - B^2 r^2)^{1/2}}$$

when $0 < x - Br < l$, and the limits of the integral will be 0 and l if $x - Br > l$. Thus, we see that although the body has no influence upstream of the characteristic cone $x = Br$ through the nose of the body,

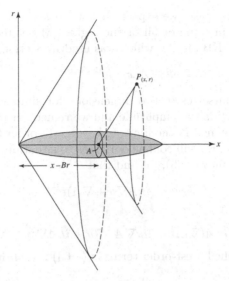

Fig. 4.9. Supersonic flow past a slender axisymmetric body.

it does affect the flow *everywhere* inside this cone, even in $x - Br > l$. This is in contrast to the result for two-dimensional supersonic flow past a thin wing where the flow is influenced by the wing only *between* the characteristics or Mach lines through the leading and trailing edges of the wing. This is an example of *Huygens' principle*, which will be discussed in more detail in Section 4.8.

4.7 High-frequency Waves

4.7.1 The Eikonal Equation

There is one other parameter regime in the frequency domain that is mathematically tractable and that is the high-frequency limit as ω or $k \to \infty$ in Helmholtz' equation. As explained at the beginning of Section 4.6, this approximation is relevant when the wavelength is small compared with the typical dimensions L of the region of interest.[18] We therefore non-dimensionalize $\mathbf{x}$ with L so that Helmholtz' equation (4.33) becomes

$$\nabla^2 \Phi + k^2 \Phi = 0, \tag{4.63}$$

where $k = \omega L/c$ is now a non-dimensional parameter which is large for light waves in many everyday situations. We can look at both interior and exterior

---

[18] Of course, if we are interested in waves sufficiently close to a sharp corner of some boundary, the method we are about to describe will never be useful, because L can then be arbitrarily small.

problems, but in any case, we expect that when $k \gg 1$, there will be *rapid spatial oscillations* in some or all of the region. Motivated by this idea, we again employ the WKB ansatz, which was used in Section 4.4.2, and write

$$\Phi \sim A(x, y)e^{iku(x,y)}, \qquad (4.64)$$

where we restrict ourselves to two dimensions for simplicity. As usual, u is the phase of Φ and A is the amplitude and we remember that these variables are only uniquely defined in the limit as $k \to \infty$. In order to save ourselves a great deal of trouble, we will confine ourselves to situations in which u is real. Substituting (4.64) into (4.63), we find that

$$\nabla \Phi \sim (ikA\nabla u + \nabla A)e^{iku}$$

and

$$\nabla^2 \Phi \sim (-k^2 A(\nabla u)^2 + 2ik\nabla A \cdot \nabla u + ikA\nabla^2 u + \nabla^2 A)e^{iku},$$

so that as $k \to \infty$, the lowest-order terms in (4.63) reveal the *eikonal equation*

$$|\nabla u|^2 = 1. \qquad (4.65)$$

The second-order terms will give the so-called "transport equation"

$$A\nabla^2 u + 2\nabla A \cdot \nabla u = 0, \qquad (4.66)$$

which determines A. Hence, we have taken the possibly retrograde step of transforming a linear second-order equation (4.63) to a fully nonlinear first-order equation (4.65), whose consideration should perhaps be deferred to Chapter 5. Assuming the reader has some familiarity with nonlinear first-order differential equations, however, we will continue here because (4.65) can give us very helpful insights into high-frequency wave propagation.

We can use *Charpit's method* (see Ockendon et al. [9]) to see that the solution of (4.66) is given by solving the characteristic equations

$$\frac{dx}{d\tau} = 2p, \qquad \frac{dy}{d\tau} = 2q, \qquad \frac{du}{d\tau} = 2, \qquad \frac{dp}{d\tau} = 0, \qquad \frac{dq}{d\tau} = 0, \qquad (4.67)$$

where $p = \partial u / \partial x$ and $q = \partial u / \partial y$, and τ parameterizes the characteristics. Initial data are required for these equations and will be given by the boundary conditions imposed on Φ, which may be at infinity for an exterior problem. We note at once that since p and q are constant from (4.67), the characteristics will be straight lines.

The simplest solution of (4.65) is $u = lx + my$, where $l^2 + m^2 = 1$, and this solution represents a plane wave. Note that the relation $l^2 + m^2 = 1$ is the dispersion relation that we found for Helmholtz' equation in Sections 4.4 and 4.5, although we are now using a different notation. More generally, it can be shown that if the high-frequency approximation to the dispersion relation for a problem is given by

$$\frac{\omega L}{c} = \Omega(\mathbf{k}L),$$

where $\mathbf{k}$ is the wavenumber as defined in Section 4.4, then the equation for u in the WKB approximation for the same problem will be

$$\Omega(\nabla(u)) = 1.$$

4.7.2 *Ray Theory

Using the terminology of Section 4.5, the nicest interpretation of the characteristics, given by (4.67), of the eikonal equation (4.65) is as *rays*. For the plane wave $u = lx + my$, the rays are straight lines in the direction (l, m), which is perpendicular to the wavefront, here defined to be the curve on which u is constant. It can easily be seen from (4.67) that this geometric relation between the rays and the wavefronts always holds.

In the case of electromagnetism, these rays are the familiar light rays that are drawn in optics in elementary physics courses. Indeed, the well-known pictures for light reflected by a mirror (Figure 4.10a) or refracted by a lens (Figure 4.10b) are easily seen to be simple superpositions of solutions of the cikonal equation. These simple pictures result from the fact that the plane

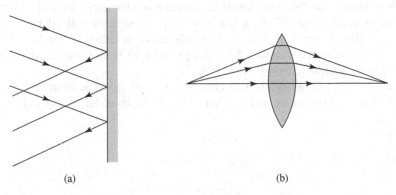

(a)   (b)

Fig. 4.10. (a) Light rays reflected from a plane mirror. (b) Light rays refracted by a lens.

wave solution of the eikonal equation is an exact solution of Helmholtz equation and so it is straightforward to see that once we have imposed suitable boundary conditions on u, the rays of the eikonal equation *reflect* in a mirror exactly as predicted in Section 4.5.2. Similarly, for *refraction* in a lens, if the speed of sound in the refractive medium is c_1, then, with $k = \omega L / c_1$, the eikonal equation is

$$|\nabla u|^2 = 1$$

in the air and

$$|\nabla u| = \left(\frac{c_0}{c_1}\right)^2$$

in the lens. When we impose continuity of u at the interface, this leads to straight-line rays in each region joined together by using Snell's law of refraction (4.36).

A much more interesting solution of (4.65) arises if we consider waves inside a circle of radius $\sqrt{2}$, say, whose perimeter is being oscillated so that

$$u = s \quad \text{on } x = \cos s + \sin s, \qquad y = \sin s - \cos s \quad \text{for } 0 < s < 2\pi, \quad (4.68)$$

and we assume k is an integer for ϕ to be 2π-periodic in s. It can be seen that on this circle, either $p = \cos s$ and $q = \sin s$, or $p = -\sin s$ and $q = \cos s$. Thus, solving (4.67) and applying these boundary conditions when $\tau = 0$ leads to two possible solutions:

(i) $u = 2\tau + s$,
       $x = 2\tau \cos s + \cos s + \sin s$,
       $y = 2\tau \sin s + \sin s - \cos s$,
       $p = \cos s$,
       $q = \sin s$;

(ii) $u = 2\tau + s$,
       $x = -2\tau \sin s + \cos s + \sin s$,
       $y = 2\tau \cos s + \sin s - \cos s$,
       $p = -\sin s$,
       $q = \cos s$.

These solutions are the two families of straight lines, shown in Figure 4.11, which intersect the circle in directions making angles $\pm\pi/4$ with the outward normal at each point. Thus, when $\tau = -\frac{1}{2}$, these rays will all touch their envelope, the circle $x^2 + y^2 = 1$, and this curve is called a *caustic*. We can verify that the transformation from (x, y) to (s, τ) is singular on the caustic since $J = \partial(x, y)/\partial(s, \tau) = 0$ when $\tau = -\frac{1}{2}$.

The fact that caustics like this can be expected to occur in most problems can be seen without using any mathematics. If we consider planar light waves

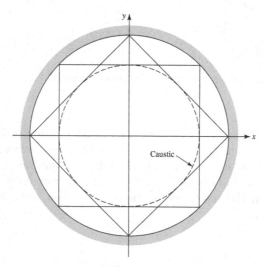

Fig. 4.11. Caustics and characteristics for the Eikonal equational with data (4.68).

with $k \gg 1$ impinging on a curved surface, then the reflected rays will always be a family of straight lines which will, in general, have an envelope. If we consider, for example, the sun shining on the inside of a circular coffee cup, the caustic which can be seen on the surface of the coffee will be a nephroid, as illustrated in Figure 4.12.

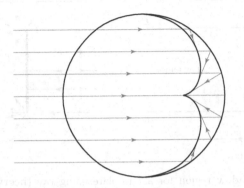

Fig. 4.12. The nephroid caustic on a cup of coffee.

In both of the above solutions, families of rays apparently stop on the caustic. Thus, the eikonal equation predicts that there will be a light region separated from a dark region by the caustic, and an analysis of the equation for A in (4.66) reveals that $|A| \to \infty$ as the caustic is approached, so that the maximum illumination is seen on the caustic itself. However, there cannot be any singularity in the solution Φ of Helmholtz' equation, which is an elliptic partial differential equation.

Before we resolve this difficulty, we mention another disturbing consequence of the fact that the eikonal equation can have real characteristics. This implies that almost any kind of singularity in the boundary data for Φ will, if k is large, cause a singularity in u to propagate away from the boundary even though Φ can have no singularities away from the boundary. However, this observation gives us the key to the success of the WKB approximation in so many situations in optics or acoustics. If we return to the diffraction problem mentioned in Section 4.5.2 for rays impinging on a flat plate, then the eikonal equation simply predicts a shadow with clear-cut edges behind the plate, as shown in Figure 4.13a. This is not a bad approximation to what happens in practice when $k \gg 1$. However, even with large k, Helmholtz' equation will predict *diffraction* at the edges A and B of the plate, and by careful analysis, it can be shown that diffracted fields propagate radially from A and B as shown schematically in Figure 4.13b. The strength of these fields in the shadow is of $O(k^{-1/2})$ relative to the incident wave field. On the other hand, the diffracted field generated by a smooth body is more subtle. Again, the high-frequency approximation predicts incident and reflected waves and a clear-cut shadow region as in Figure 4.14a, but now it can be shown that the amplitude of the field in the shadow region is much smaller being of $O(e^{-k^{1/3}})$ compared to the incident field. The shadow indicated in Figure 4.14b needs to be described by rays which are known as *creeping rays* rather than the diffracted rays of Figure 4.13b (see Born and Wolf [15]).

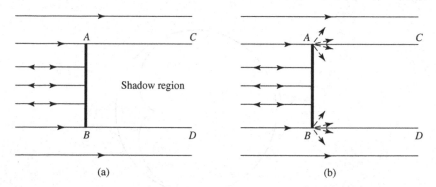

Fig. 4.13. (a) Shadow region for a flat plate using ray theory. (b) The effect of diffraction at the edges of a flat plate.

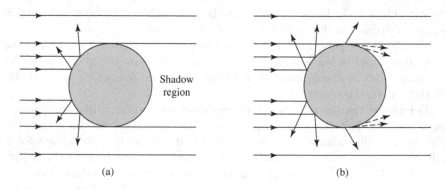

Fig. 4.14. (a) Rays reflected from a smooth body using ray theory. (b) The effect of diffraction for a smooth body.

Both of the shadow regions can be described by real ray theory if we are prepared to undertake all of the complicated "singular perturbation" analysis which is needed to unravel the structure of the solution near the diffraction points A and B. More importantly, with even more work, thin layers, analogous to boundary layers, can be constructed along AC and BD to smooth out the discontinuities between the incoming waves and the refracted field. This not only resolves the apparent contradiction between the hyperbolicity or the eikonal equation and the ellipticity of Helmholtz' equation, but it also gives us the clue as to what happens in the even deeper shadow within the central circle in Figure 4.12, where the wave field appears to end in a caustic. In this case, a boundary layer analysis close to the caustic reveals that the field inside the caustic is of $O(e^{-k})$ as $k \to \infty$ and that the rays in this region are *complex* and x, y, p, q, and u all take complex values. Unfortunately, complex ray theory requires an even more elaborate asymptotic development than does real ray theory (see Chapman et al. [16]).

We remark that if we solve the eikonal equation with the boundary condition (4.68) replaced by $u = \varepsilon s$ on the same circle, where $k\varepsilon$ is now an integer, we find that the caustic lies closer and closer to the boundary $x^2 + y^2 = 2$ as $\varepsilon \to 0$. Now, although the introduction of a boundary layer structure can smooth out the discontinuity on the caustic, most of the wave field will still be "trapped" close to the boundary, and this is an example of the famous *whispering gallery waves* that can easily be generated inside the dome of St. Paul's Cathedral, for example. Another readily observed caustic, but one with quite a different structure, is the boundary $\sin \theta = \pm 1/2\sqrt{2}$ of the ship wave pattern described in Section 4.6.1.

4.8 *Dimensionality and the Wave Equation

At several points in this chapter, we have remarked on the way in which the *qualitative* nature of linear wave propagation as described by the wave equation (3.6) depends on the number of space dimensions. The most striking piece of evidence for this has been the observation in Sections 4.6.3 and 4.6.4 that a thin wing moving supersonically in two dimensions leaves behind it no wake at all, whereas a slender axisymmetric projectile can always be detected after its passage.

A more familiar scenario concerns the propagation of a disturbance that is localized near a point in space and time. We know that for the one-dimensional case of, say, waves on a string, an initial disturbance localized near $x = 0$ will eventually emerge as two pulses, each propagating without change or diminution near $x = \pm ct$, and there will be no disturbance anywhere else. Similarly, using the solution (4.5) in three dimensions, an initial disturbance near $r = 0$ will evolve into one that is localized in the space (x, y, z, t) near the spherical "shell" $x^2 + y^2 + z^2 = c^2 t^2$, albeit with a decay factor of r^{-1}. This is in accord with the evidence from our eardrums, but when we look at a disturbance initially localized on the surface of a drum or created by dropping a pebble into a pond, the situation is quite different.

To appreciate this difference, we first remark that the localization of the waves in the string example mentioned earlier is identical to the localization of the waves emitted by the supersonic wing as given by (4.53) and (4.54) in the limit $l \to 0$. We simply have to identify x with time and y with distance along the string to make the problems mathematically equivalent. Hence, we are in a position to understand the evolution of two-dimensional axisymmetric waves satisfying

$$\frac{\partial^2 \phi}{\partial r^2} + \frac{1}{r}\frac{\partial \phi}{\partial r} = \frac{1}{c^2}\frac{\partial^2 \phi}{\partial t^2} \tag{4.69}$$

by drawing a similar analogy with the steady axisymmetric supersonic flow problem described in Section 4.6.4 when $M > 1$. When we again let the length of the body tend to zero, which is equivalent to the release of a short pulse of sound at $t = 0$ in (4.69), we find that the solution (4.60) gives us

$$\phi = \begin{cases} 0, & r > ct \\ \dfrac{\lambda}{(c^2 t^2 - r^2)^{1/2}}, & r < ct, \end{cases} \tag{4.70}$$

for some constant λ. Hence, although there is a sharp front at $r = ct$ as in the one- and three-dimensional cases, in two dimensions the disturbance is felt everywhere inside the cone $r = ct$ and is *not* localized near the cone $r = ct$. Those who doubt this argument may ask themselves why a lightning strike, which may be approximated as an instantaneous energy release along a vertical line, produces *rumbles* of thunder after the sharp crack at $r = ct$.

The mathematics can be beautifully unified by the theory of the *retarded potential*. It can be shown (Ockendon et al. [9]) that the solution of the three-dimensional wave equation with an initial distribution $\partial \phi / \partial t = f(x, y, z)$, $\phi = 0$ is given by the formula

$$\phi(x, y, z, t)$$
$$= \frac{t}{4\pi}\int_0^{2\pi}\int_0^{\pi} f(x + ct\sin\theta\cos\phi, y + ct\sin\theta\sin\phi, z + ct\cos\theta)\sin\theta\, d\theta\, d\phi. \tag{4.71}$$

It can be verified (after much work) that this function satisfies the wave equation and, in fact, a linear combination of ϕ and $\partial \phi / \partial t$ can be used to satisfy any given initial conditions. Now, the physical significance of (4.71) is that if f is zero outside a region D, then the integral will be non-zero only for times $t_{\min} < t < t_{\max}$ where, as shown in Figure 4.15, $t_{\min}$ is the first time the disturbance is felt at (x, y, z), and $t_{\max}$ is the last time.

We see at once that if the initial disturbance is very elongated in the z direction say, then $t_{\max}$ will tend to infinity while $t_{\min}$ remains fixed. Hence we have *Huygens' principle* that there are sharp leading and trailing wave fronts in one or three dimensions, whereas in two dimensions, there is only a sharp "leading" wave front.

We must emphasize that the above discussion only applies to the wave equation with constant coefficients. For waves in inhomogeneous media (including three-dimensional waves in a half-space) or for waves governed by models other than (3.6), there is no reason for Huygens' principle to hold. What is true generally is that the waves decay with distance more rapidly as the number of space dimensions increases. However, it is no easy matter to estimate this rate of decay for a general wave model.

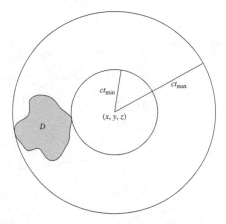

Fig. 4.15. The effect of an internal disturbance in D.

The above discussion can give us insight into the possible evolution of the steady wave patterns considered in Sections 4.6.4 and 4.6.5. The expression (4.71) is the general solution of

$$\frac{\partial^2 \phi}{\partial x^2} + \frac{\partial^2 \phi}{\partial y^2} + \frac{\partial^2 \phi}{\partial z^2} = \frac{1}{c^2}\frac{\partial^2 \phi}{\partial t^2},$$

which is sometimes called the wave equation in the *acoustic frame*. However, in Section 4.6.4, we set $\xi = x - Ut$ and considered solutions of the wave equation in the *aerodynamic frame*, namely

$$\left(1 - \frac{U^2}{c^2}\right)\frac{\partial^2 \phi}{\partial \xi^2} + \frac{\partial^2 \phi}{\partial y^2} + \frac{\partial^2 \phi}{\partial z^2} = \frac{1}{c^2}\frac{\partial^2 \phi}{\partial t^2} - \frac{2U}{c^2}\frac{\partial^2 \phi}{\partial \xi \partial t},$$

and we restricted ourselves to the "steady" case where $\phi = \phi(\xi, y, z)$.

To see how such a steady flow might be set up, let us consider, in the acoustic frame, what happens when a two-dimensional disturbance is localized near $x = Ut$ and $y = 0$ for $t \geq 0$ and with $U > c$. The solution at time t consists of the superposition of the solutions generated at times τ, where $0 < \tau < t$, and the above discussion reveals that the contribution from time τ is contained within the cone $(x - U\tau)^2 + y^2 = c^2(t - \tau)^2$ in (x, y, t) space. From (4.70), the amplitude of this contribution is non-zero inside this cone and is infinite at the cone surface, which is sometimes called the *wavefront* (yet another use of the term). Hence, the solution at time t in the acoustic frame is contained within the superposition of the cones as shown in Figure 4.16a. The projection of the "tops" of the cones in the aerodynamic (x, y) plane is simply the sequence of circles shown in figure 4.16b beginning from the "starting" circle $x^2 + y^2 = c^2 t^2$. We thus see the following:

(i) The characteristics in the aerodynamic frame, namely $\xi = \pm y\sqrt{U^2/c^2 - 1}$, emerge as the *envelope* of the wave fronts in the acoustic frame.

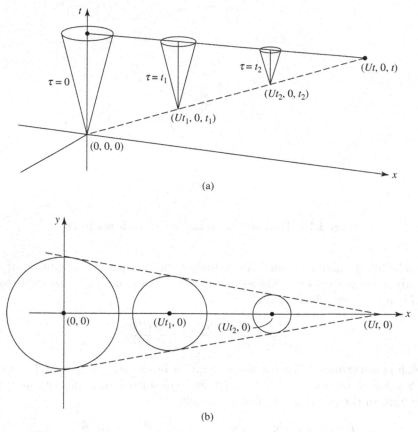

Fig. 4.16. (a) The influence of a supersonic point source of sound at $x = Ut$, $y = 0$ in (x, y, t) space. (b) The influence of a supersonic point source in (x, y) space.

(ii) There is a complicated non-zero disturbance between these characteristics for all finite time. However, as $t \to \infty$, this disturbance decays through destructive interference, leaving the "no-wake" flow described in Section 4.6.4.

Exercises

4.1 (i) It can be shown that, for steady small two-dimensional disturbances on a uniform flow in a weakly stratified gas with Mach number M, the velocity potential ϕ of the disturbance satisfies

$$M^2 \frac{\partial^2 \phi}{\partial x^2} = \frac{\partial^2 \phi}{\partial x^2} + \frac{\partial^2 \phi}{\partial y^2}.$$

If $M = (1 + y)^{1/2}$, in which case the equation is called a *Tricomi equation*, show that the characteristics are

$$\pm 2y^{3/2} = 3x + \text{constant}.$$

Sketch these characteristics and indicate the "region of influence" of a small obstacle which is put in the flow at a point where $y > 0$.

(ii) Suppose that, in the same stratified flow, a wave of the form $\phi = \text{Rl}(e^{ikx}A(y))$ is incident on $y > 0$ from above. Show that

$$\frac{d^2 A}{dy^2} + k^2 y A = 0.$$

This is *Airy's equation* and it shows that ϕ will be oscillatory in $y > 0$, but that, in $y < 0$, it will decay exponentially. Thus, waves from above will not penetrate far into $y < 0$ and the disturbance due to the obstacle in part (i) will decay exponentially in $y < 0$.

R4.2 In axisymmetric acoustic wave propagation inside a rigid circular cylinder of radius a, the velocity potential $\phi(r, t)$ satisfies

$$\frac{1}{c^2}\frac{\partial^2 \phi}{\partial t^2} = \frac{\partial^2 \phi}{\partial r^2} + \frac{1}{r}\frac{\partial \phi}{\partial r} \quad \text{for } 0 \le r < a,$$

with

$$\frac{\partial \phi}{\partial r} = 0 \quad \text{on } r = a.$$

Show that ϕ can be written as a generalized Fourier series in the form

$$\phi = \sum_{n=0}^{\infty}(a_n \cos \omega_n t + b_n \sin \omega_n t)J_0\left(\frac{\omega_n r}{c}\right),$$

where $J_0(z)$ satisfies

$$\frac{d^2 J_0}{dz^2} + \frac{1}{z}\frac{dJ_0}{dz} + J_0 = 0, \tag{$\dagger$}$$

with $J_0(0) = 1$, ω_n defined by $J_0'(\omega_n a/c) = 0$, and a_n and b_n arbitrary constants. It can be shown[19] that

$$\int_0^a rJ_0\left(\frac{\omega_n r}{c}\right)J_0\left(\frac{\omega_m r}{c}\right)dr = \begin{cases} 0, & m \ne n \\ \frac{1}{2}a^2\left[J_0\left(\frac{\omega_n a}{c}\right)\right]^2, & m = n. \end{cases}$$

Show that if the gas in the tube is initially at rest with a pressure distribution $P_0(r)$, then

$$\phi(r,0) = 0 \quad \text{and} \quad \frac{\partial \phi}{\partial t}(r,0) = -\frac{1}{\rho_0}P_0(r).$$

---

[19] To prove this orthogonality result, put $z = \omega_n r/c$ in ($\dagger$) and then multiply by $2r^2 J_0'(\omega_n r/c)$ and integrate from $r = 0$ to a, integrating by parts once.

Hence, show that $a_n = 0$ and that

$$b_n = \frac{-2}{\rho_0 a^2 \omega_n} \frac{\int_0^a r P_0(r) J_0(\omega_n r/c) dr}{[J_0(\omega_n a/c)]^2}.$$

R4.3 Consider one-dimensional acoustic waves in an organ pipe which is closed at $x = 0$ and open to the atmosphere at $x = L$. Assuming constant pressure at $x = L$, show that the velocity potential ϕ satisfies

$$\frac{\partial \phi}{\partial x}(0, t) = 0 \quad \text{and} \quad \phi(L, t) = 0.$$

If $\phi(x, 0) = f(x)$ and $\partial \phi(x, 0)/\partial t = g(x)$, show that

$$\phi = \sum_{n=0}^{\infty} \left(a_n \cos \frac{(2n+1)\pi ct}{2L} + b_n \sin \frac{(2n+1)\pi ct}{2L} \right) \cos \frac{(2n+1)\pi x}{2L},$$

where

$$a_n = \frac{2}{L} \int_0^L f(x) \cos \frac{(2n+1)\pi x}{2L} dx$$

and

$$b_n = \frac{4}{(2n+1)\pi c} \int_0^L g(x) \cos \frac{(2n+1)\pi x}{2L} dx.$$

This could be a model for a motor bicycle exhaust; so, suppose, more realistically, that the open end radiates sound to the environment. One model for this is to say that at $x = L$,

$$\frac{\partial \phi}{\partial t} + \alpha \frac{\partial \phi}{\partial x} = 0.$$

Show that if we assume that high pressure in the pipe pumps gas into the environment and low pressure sucks gas into the pipe, then $\alpha > 0$.

By separating the variables in (4.1), show that

$$\phi = (A \cos \omega_n t + B \sin \omega_n t) \cos \frac{\omega_n x}{c},$$

where

$$\tan^2 \frac{\omega_n L}{c} + \frac{c^2}{\alpha^2} = 0.$$

Deduce that ω_n is complex and show that for $\alpha \ll c$,

$$\omega_n = \frac{(2n+1)\pi c}{2L} \pm i \frac{\alpha}{L}.$$

Show that the solution of the initial value problem is now

$$\phi \simeq \sum_{n=0}^{\infty} e^{-\alpha t/L} \left(a_n \cos \frac{(2n+1)\pi ct}{2L} + b_n \sin \frac{(2n+1)\pi ct}{2L} \right) \cos \frac{(2n+1)\pi x}{2L},$$

where a_n and b_n are as given above.

*4.4 Show that the Fourier transform of $\varepsilon/(x^2 + \varepsilon^2)$ is $\pi e^{-\varepsilon|k|}$. If the impact of a long stick on a pond is modeled by taking the initial surface profile to be $\eta_0(x) = -a\varepsilon/\pi(x^2 + \varepsilon^2)$, show that

$$\eta(x,t) = -\frac{a}{\pi} \int_0^\infty e^{-\varepsilon k} \cos\sqrt{gk}t \cos kx \, dk.$$

Use the method of stationary phase to show that as $x, t \to \infty$, the major contribution to the integral comes from values of k satisfying

$$x \pm \frac{1}{2}\sqrt{\frac{g}{k}}t = 0,$$

and hence that for $x > 0$,

$$\eta \sim -\frac{at}{2}\sqrt{\frac{g}{\pi x^3}} \cos\left(\frac{gt^2}{4x} - \frac{\pi}{4}\right)$$

when ε is sufficiently small.

Which of the pictures in Figure 4.17 is the more realistic representation of this long-time behavior?

(a)   (b)

Fig. 4.17. Possible waves generated by a localised source.

R4.5 The effect of a constant surface tension T on two-dimensional interfacial gravity waves is to introduce a pressure drop of T/R across the interface $y = \eta(x,t)$, where R is the local radius of curvature of the interface. Show that, with a sign convention that you should specify, R^{-1} is approximately $\partial^2\eta/\partial x^2$ for small-amplitude waves.

Show that if a liquid of density ρ_1, pressure p_1 lies above a liquid of density ρ_2 and pressure p_2, with the interface being given by $z = \eta(x,t)$, then

$$p_1 - p_2 = T\frac{\partial^2\eta}{\partial x^2} \quad \text{on } z = 0,$$

to lowest order. Deduce that if $\eta = a\cos(kx - \omega t)$, the dispersion relation is

$$\omega^2 = g|k|\frac{(\rho_2 - \rho_1)}{(\rho_2 + \rho_1)} + \frac{T|k|^3}{(\rho_2 + \rho_1)}.$$

This implies that surface tension can stabilize such an interface even if $\rho_1 > \rho_2$. Is surface tension more effective as a stabilizing mechanism for large $|k|$ or for small $|k|$?

*4.6 The displacement $\mathbf{u}$ in an elastic medium satisfies (3.33). If $\mathbf{u}$ is written as $\mathbf{u} = \mathrm{Rl}\,\mathbf{u}_0 e^{i(\mathbf{k}\cdot\mathbf{x}-\omega t)}$, where $\mathbf{u}_0 = A\mathbf{k} + \mathbf{B}\wedge\mathbf{k}$, with $A, \mathbf{k},$ and $\mathbf{B}$ all constant, show that

$$\text{either} \quad \mathbf{B} = 0 \quad \text{and} \quad (\lambda + 2\mu)|k|^2 = \rho\omega^2,$$
$$\text{or} \quad A = 0 \quad \text{and} \quad \mu|k|^2 = \rho\omega^2.$$

If $k_p = \omega\sqrt{\rho/(\lambda + 2\mu)}$ and $k_s = \omega\sqrt{\rho/\mu}$, and c_p and c_s are the corresponding wave speeds, it can be shown that $c_p > c_s$.

The boundary conditions for a two-dimensional displacement $\mathbf{u} = (u(x,y,t), v(x,y,t), 0)$ in $y < 0$, with an unstressed boundary at $y = 0$, are

$$\frac{\partial u}{\partial y} + \frac{\partial v}{\partial x} = 0 = 2\mu\frac{\partial v}{\partial y} + \lambda\left(\frac{\partial u}{\partial x} + \frac{\partial v}{\partial y}\right)$$

(see Exercise 3.12). Show that a solution for a propagating wave in the form

$$\mathbf{u} = \mathrm{Rl}(\mathbf{a}_p e^{\kappa_p y} + \mathbf{a}_s e^{\kappa_s y})e^{i(kx-\omega t)},$$

where $\kappa_p^2 = k^2 - k_p^2$ and $\kappa_s^2 = k^2 - k_s^2$, is possible as long as

$$\left(2 - \frac{c^2}{c_s^2}\right)^2 = 4\left(1 - \frac{c^2}{c_p^2}\right)^{1/2}\left(1 - \frac{c^2}{c_s^2}\right)^{1/2},$$

where $c = \omega/k < c_s$. This wave is called a *Rayleigh wave*.

*4.7 From Exercise 3.1, the perturbation potential ϕ for small two-dimensional disturbances in a uniform flow $U\mathbf{i}$ satisfies

$$\left(1 - \frac{U^2}{c^2}\right)\frac{\partial^2\phi}{\partial x^2} + \frac{\partial^2\phi}{\partial y^2} = \frac{2U}{c^2}\frac{\partial^2\phi}{\partial x\partial t} + \frac{1}{c^2}\frac{\partial^2\phi}{\partial t^2},$$

where c is the speed of sound in the undisturbed gas. The gas is in $y > 0$ and flows past an elastic membrane under tension T which originally lies on $y = 0$ and is subject to small displacements $y = \eta(x,t)$. The boundary conditions on the membrane are that on $y = 0$,

$$\frac{\partial\phi}{\partial y} = \frac{\partial\eta}{\partial t} + U\frac{\partial\eta}{\partial x}$$

and

$$\frac{p}{\rho_0} = \frac{T}{\rho_0}\left(\frac{\partial^2\eta}{\partial x^2} - \frac{1}{c_m^2}\frac{\partial^2\eta}{\partial t^2}\right) = -\left(\frac{\partial\phi}{\partial t} + U\frac{\partial\phi}{\partial x}\right),$$

where c_m^2 is the speed of sound in the membrane and ρ_0 is the ambient density in the gas.

By writing

$$\phi = \mathrm{Rl}(Ae^{i(kx-\omega t)-\lambda y}), \qquad \eta = \mathrm{Rl}(ae^{i(kx-\omega t)})$$

with $\text{Rl}\,\lambda \geq 0$, deduce that

$$k^2 - \lambda^2 = \left(\frac{Uk}{c} - \frac{\omega}{c}\right)^2 \quad \text{and} \quad \frac{\lambda T}{\rho_0}\left(k^2 - \frac{\omega^2}{c_m^2}\right) = (Uk - \omega)^2.$$

Deduce that, as $\rho_0 \to 0$ with k real, ω is complex and the motion is unstable if $U > c + c_m$.

This is an example of "flutter," a phenomenon from which aerodynamic surfaces can suffer.

*4.8 (i) Helmholtz' equation in cylindrical polar coordinates (r, θ) is

$$\frac{\partial^2 \Phi}{\partial r^2} + \frac{1}{r}\frac{\partial \Phi}{\partial r} + \frac{1}{r^2}\frac{\partial^2 \Phi}{\partial \theta^2} + k^2 \Phi = 0.$$

To study the far field, write $r = R/\varepsilon$, where $\varepsilon \ll 1$, and use the ansatz $\Phi \sim A e^{iku/\varepsilon}$ to show that

$$\left(\frac{\partial u}{\partial R}\right)^2 = 1$$

and

$$2\frac{\partial A}{\partial R} + \frac{A}{R} = 0.$$

Deduce that when waves propagate outward radially for large r,

$$\Phi \sim \frac{A_0(\theta)}{r^{1/2}}e^{ikr},$$

and hence that $r^{1/2}(\partial \Phi/\partial r - ik\Phi) \to 0$ as $r \to \infty$.

(ii) Helmholtz' equation in spherical polar coordinates (r, θ, ψ) is

$$\frac{\partial^2 \Phi}{\partial r^2} + \frac{2}{r}\frac{\partial \Phi}{\partial r} + \frac{1}{r^2 \sin\theta}\frac{\partial}{\partial \theta}\left(\sin\theta \frac{\partial \Phi}{\partial \theta}\right) + \frac{1}{r^2 \sin\theta}\frac{\partial^2 \Phi}{\partial \psi^2} + k^2 \Phi = 0.$$

Repeat the scaling in (i) to show that for large r,

$$\Phi \sim \frac{A_0(\theta, \psi)}{r}e^{ikr}$$

and hence that $r(\partial \Phi/\partial r - ik\Phi) \to 0$ as $r \to \infty$.

R4.9 In the frequency domain, ϕ_1 and ϕ_2 satisfy

$$(\nabla^2 + k_1^2)\phi_1 = 0 \quad \text{in } y > 0$$

and

$$(\nabla^2 + k_2^2)\phi_2 = 0 \quad \text{in } y < 0,$$

and a class of refraction problems leads to the conditions that $k^2\phi$ and $\partial \phi/\partial y$ are continuous at $y = 0$.

Show that if $y = 0$ is irradiated from above by a plane wave

$$\phi_i = A e^{ik_1(y \cos \theta_1 - x \sin \theta_1)},$$

so that the incident rays are in the $(-\sin \theta_1, \cos \theta_1)$ direction, then

$$\phi_1 = \phi_i + R e^{-ik_1(y \cos \theta_1 + x \sin \theta_1)}$$

and

$$\phi_2 = T e^{ik_2(y \cos \theta_2 - x \sin \theta_2)},$$

where $k_1 \sin \theta_1 = k_2 \sin \theta_2$.

As long as $\lambda = k_1 \sin \theta_1 / k_2 < 1$, this "conservation of tangential wavenumber" is *Snell's law of refraction*. Since $k_i = \omega/c_i$, it implies that $\sin \theta_1 / \sin \theta_2 = c_1/c_2$. The same law holds for optics (see Billingham and King [17] for an analysis of Maxwell's equations) and since the speed of light in water is less than the speed of light in air, Snell's law shows that light is bent towards the normal on entering water, making ponds seem shallower than they really are.

If $k_1 > k_2$, it is possible for λ to be greater than unity, and in this case there is "*total internal reflection.*" Show that if $\lambda > 1$,

$$\phi_2 = B e^{k_2 \sqrt{\lambda^2 - 1}\, y - i\lambda k_2 x}$$

and that, in this case, $|R| = |A|$.

4.10 Elastic waves in a plate can be modeled by the equation

$$D \left(\frac{\partial^2}{\partial x^2} + \frac{\partial^2}{\partial y^2} \right)^2 u + \rho \frac{\partial^2 u}{\partial t^2} = 0,$$

where $u(x, y)$ is the normal displacement of the plate and D is a positive constant proportional to the Young's modulus of the material from which the plate is made. Show that the dispersion relation is

$$\rho \omega^2 = D(k_1^2 + k_2^2)^2.$$

In one dimension, the model reduces to the *beam equation*

$$D \frac{\partial^4 u}{\partial x^4} + \rho \frac{\partial^2 u}{\partial t^2} = 0$$

and suitable boundary conditions for a beam which is clamped at both ends are $u = \partial u/\partial x = 0$ at the ends. Show that the resonant frequencies for a clamped beam are not rationally related to each other.

Note that this result is in contrast to solutions for the transverse vibrations of a string for which

$$-T \frac{\partial^2 u}{\partial x^2} + \rho \frac{\partial^2 u}{\partial t^2} = 0$$

with $u = 0$ at $x = 0$ and l, and the resonant frequencies are $(n\pi/l)\sqrt{(T/\rho)}$.

*4.11 (i) The frequency domain model for one-dimensional waves in a periodic medium in which the inhomogeneity is weak is the Matthieu equation

$$\frac{d^2\Phi}{dx^2} + (k^2 + \varepsilon \cos 2x)\Phi = 0,$$

where $k = O(1)$ and $\varepsilon \ll 1$. Show that a perturbation solution

$$\Phi \sim A \cos kx + B \sin kx + \varepsilon \phi_1 + \varepsilon^2 \phi_2 + \cdots$$

reveals that the terms $\varepsilon \phi_1$, $\varepsilon^2 \phi_2$, $\varepsilon^3 \phi_3, \ldots$ cannot all remain small compared to the lowest-order term for all x if k is an integer.

Suppose that $k^2 = 1 + \kappa \varepsilon$, where κ is $O(1)$. Show that

$$\frac{d^2\phi_1}{dx^2} + \phi_1 = -\kappa(A \cos x + B \sin x)$$

$$-\frac{1}{2}(A(\cos 3x + \cos x) + B(\sin 3x - \sin x)).$$

Deduce that Φ can only be periodic to order ε if $\kappa = \frac{1}{2}$ and $A = 0$, or if $\kappa = -\frac{1}{2}$ and $B = 0$.

(ii) If $k^2 = 1 + \kappa \varepsilon$, as above, but x is large so that $x = X/\varepsilon$, where $X = O(1)$, show that the WKB solution of the equation is

$$\Phi \sim A e^{iX/\varepsilon} + A^* e^{-iX/\varepsilon},$$

where

$$2i\frac{dA}{dX} + \kappa A + \frac{1}{2}A^* = 0 \quad \text{and} \quad -2i\frac{dA^*}{dX} + \kappa A^* + \frac{1}{2}A = 0.$$

Deduce that A and A^* grow or decay exponentially in X if $|\kappa| < \frac{1}{2}$ and that they are oscillatory in X if $|\kappa| > \frac{1}{2}$.

This example shows that waves in such a periodic medium will decay if k is in the *stop band*, which is $-\varepsilon/4 < k - 1 < \varepsilon/4$ approximately. It can be shown that similar stop bands exist near any integer value of k (including 0) as shown in Figure 4.5. Hence, over large regions of the (k, ε) parameter space, the material acts to damp waves exponentially rather than allowing them to propagate. This is an example of *Floquet theory*, which, in higher dimensions, is associated with the names of Bloch and Brillouin.

4.12 A one-dimensional acoustic resonator is closed at $x = 0$ and is driven at $x = 1$ by a piston which oscillates with a frequency which is much lower than any of the resonant frequencies of the pipe. Show that a suitable non-dimensional model is

$$\frac{\partial^2 \phi}{\partial x^2} = \frac{\partial^2 \phi}{\partial t^2} \quad \text{for } 0 < x < 1,$$

with $\partial \phi / \partial x = 0$ at $x = 0$ and $\partial \phi / \partial x = \sin \varepsilon t$ at $x = 1$ where $\varepsilon \ll 1$. Assuming a periodic response, solve this problem and show that as $\varepsilon \to 0$, the pressure response has amplitude of $O(\varepsilon^{-1})$, but that the gas velocity will never be greater than its maximum value at the piston.

This is a very simple example of a *Helmholtz resonator*, which is driven at a frequency very much less than any of the natural frequencies.

4.13 (i) Suppose that N is constant in the inertial wave model (3.26) and that $w = \text{Rl}(W(x, y, z)e^{-i\omega t})$. Show that if g is sufficiently large, then

$$(N^2 - \omega^2)\left(\frac{\partial^2 W}{\partial x^2} + \frac{\partial^2 W}{\partial y^2}\right) = \omega^2 \frac{\partial^2 W}{\partial z^2}$$

and deduce that waves can radiate to infinity if $\omega^2 < N^2$.

N is called the *Brunt-Väisälä frequency*.

(ii) Show from (3.30) that if $p = \text{Rl}(P(x, y, z)e^{-i\omega t})$ in steady rotating flow at a high Rossby number, then P satisfies

$$\frac{\partial^2 P}{\partial x^2} + \frac{\partial^2 P}{\partial y^2} + \frac{\partial^2 P}{\partial z^2} = \frac{4\Omega^2}{\omega^2}\frac{\partial^2 P}{\partial z^2}.$$

Show that waves can radiate to infinity if $\omega^2 < 4\Omega^2$.

Note that in both these cases, there is a *cut-off frequency* above which waves cannot radiate to infinity.

4.14 Show that the wave crests and troughs in the far-field ship's wave pattern (4.39) are given by the curves $k_1 x + k_2 y = c$, where k_1 and k_2 are related by (4.40) and c is a constant. Remembering that $y = \lambda x$, where λ is given by (4.42), and putting $U^4/g^2 = 1$ for simplicity, show that the crests and troughs are given parametrically in terms of k_1 by

$$x = \frac{c(2k_1^2 - 1)}{k_1^3}, \qquad y = \mp\frac{(k_1^2 - 1)^{1/2}c}{k_1^3}.$$

Show that for a fixed c, y is maximum when $k_1^2 = 3/2$ and $y/x = \mp 1/2\sqrt{2}$. Show that $dy/dx = \mp 1/(k_1^2 - 1)^{1/2}$ and sketch these curves for fixed c. Hence, show that the crests in a ship's wake consist of two families of curves as shown in Figure 4.6.

*4.15 Steady two-dimensional homentropic flow in a slender nozzle is modeled by (2.6)–(2.8) in the form

$$\frac{\partial}{\partial x}(\rho u) + \frac{\partial}{\partial y}(\rho v) = 0,$$

$$\rho\left(u\frac{\partial u}{\partial x} + v\frac{\partial u}{\partial y}\right) = -\frac{\partial p}{\partial x}$$

and

$$\rho\left(u\frac{\partial v}{\partial x} + v\frac{\partial v}{\partial y}\right) = -\frac{\partial p}{\partial y}$$

with

$$\frac{v}{u} = \pm\frac{\varepsilon^2}{2}\bar{A}' \quad \text{on } y = \pm\frac{1}{2}(A_0 + \varepsilon\bar{A}(\varepsilon x)),$$

where $A_0 = $ constant. Show that if $\varepsilon x = X$, $u = u_0 + \varepsilon u'$, $v = \varepsilon^2 v'$, $p = p_0 + \varepsilon p'$, and $\rho = \rho_0 + \varepsilon \rho'$, where u_0, p_0, and ρ_0 are constant, then, to lowest order,

$$\rho_0 \frac{\partial u'}{\partial X} + u_0 \frac{\partial \rho'}{\partial X} + \rho_0 \frac{\partial v'}{\partial y} = 0,$$

$$\rho_0 u_0 \frac{\partial u'}{\partial X} = -\frac{\partial p'}{\partial X}, \qquad 0 = \frac{\partial p'}{\partial y}$$

and

$$p' = \frac{\gamma p_0}{\rho_0} \rho',$$

with

$$v' = \pm \frac{u_0}{2} \frac{d\bar{A}}{dX} \qquad \text{on } y = \pm \frac{A_0}{2}.$$

Deduce that, if $\bar{u} = 1/A_0 \int_{-A_0/2}^{A_0/2} u' \, dy$, and similarly for $\bar{p}$ and $\bar{\rho}$, then

$$A_0 \rho_0 \frac{d\bar{u}}{dX} + A_0 u_0 \frac{d\bar{\rho}}{dX} + \rho_0 u_0 \frac{d\bar{A}}{dX} = 0,$$

$$\rho_0 u_0 \frac{d\bar{u}}{dX} + \frac{d\bar{p}}{dX} = 0$$

and

$$\bar{p} = \frac{\gamma p_0}{\rho_0} \bar{\rho}.$$

Hence show that $\bar{\rho} = \rho_0 \bar{A} M^2 / A_0 (1 - M^2)$, where $M^2 = u_0^2 \rho_0 / \gamma p_0$.

*4.16 (i) Show that if $\phi(x, y)$ is given by (4.50), then

$$\frac{\partial \phi}{\partial y} = 2\beta^2 y \int_0^l \frac{g(\xi) \, d\xi}{(x - \xi)^2 + \beta^2 y^2}.$$

Noting that, as $y \to 0$, the major contribution to the integral comes from near $\xi = x$, show that

$$\lim_{y \downarrow 0} \frac{\partial \phi}{\partial y} = \begin{cases} 2\pi \beta g(x), & 0 < x < l \\ 0, & \text{otherwise} \end{cases}$$

and

$$\lim_{y \uparrow 0} \frac{\partial \phi}{\partial y} = \begin{cases} -2\pi \beta g(x), & 0 < x < l \\ 0, & \text{otherwise.} \end{cases}$$

Use the same type of argument to show that the integral in (4.58) tends to $h(x)[\sinh^{-1}(l - x/\beta r) + \sinh^{-1} x/\beta r]$ as $r \to 0$ and deduce that $\phi \sim -2h(x) \log r$ in this limit.

(ii) Show that the potential

$$\phi(x, y) = \int_0^l h(\xi) \tan^{-1} \left(\frac{x - \xi}{\beta y} \right) d\xi \qquad (*)$$

satisfies (4.49), where the function $\tan^{-1}$ is defined to lie between $-\pi/2$ and $\pi/2$. Show that as $y \downarrow 0$,

$$\tan^{-1}\left(\frac{x-\xi}{\beta y}\right) \to \pm\frac{\pi}{2} - \frac{\beta y}{x-\xi} + O(y^2) \qquad (**)$$

according to whether $x-\xi > 0$ or $x-\xi < 0$ respectively. Show directly from (*) that, for $0 < x < l$ with y fixed,

$$\phi(x,y) = \lim_{\varepsilon \to 0}\left(\int_0^{x-\varepsilon} + \int_{x+\varepsilon}^l\right) h(\xi)\tan^{-1}\left(\frac{x-\xi}{\beta y}\right)\,d\xi.$$

Hence, use (**) to show that for small positive values of y, ϕ is approximately

$$\int_0^{x-\varepsilon} h(\xi)\left(\frac{\pi}{2} - \frac{\beta y}{x-\xi}\right)\,d\xi + \int_{x+\varepsilon}^l h(\xi)\left(-\frac{\pi}{2} - \frac{\beta y}{x-\xi}\right)\,d\xi.$$

Deduce that as $y \downarrow 0$,

$$\frac{\partial\phi}{\partial y} \to -\beta \fint_0^l \frac{h(\xi)}{x-\xi}\,d\xi,$$

where the *Cauchy principal value* integral is defined by

$$\fint_0^l \frac{h(\xi)}{x-\xi}\,d\xi = \lim_{\varepsilon \to 0}\left(\int_0^{x-\varepsilon} + \int_{x+\varepsilon}^l\right)\frac{h(\xi)}{x-\xi}\,d\xi.$$

Show that $\partial\phi/\partial y$ takes the same value as $y \uparrow 0$ and deduce that the problem for subsonic flow past an infinitely thin *asymmetric* aerofoil $y = \varepsilon f_A(x)$ requires us to solve the *singular integral equation*

$$-\frac{U}{\beta}f_A'(x) = \fint_0^l \frac{h(\xi)\,d\xi}{x-\xi}.$$

Note that an arbitrary wing shape $\varepsilon f_\pm(x)$ can always be written as $\varepsilon(\pm f_S + f_A)$, where $f_S = \frac{1}{2}(f_+ - f_-)$ and $f_A = \frac{1}{2}(f_+ + f_-)$.

4.17 Solve the problem of subsonic flow past a symmetric thin wing by using Fourier transforms as follows. If $\bar\phi$ is defined by $\bar\phi = \int_{-\infty}^{\infty} \phi(x,y)e^{ikx}\,dx$, show that (4.49) and (4.48) lead to the problem

$$\frac{d^2\bar\phi}{dy^2} - \beta^2 k^2 \bar\phi = 0 \quad \text{in } y > 0,$$

with

$$\frac{d\bar\phi}{dy} = U\bar F(k) \quad \text{on } y = 0,$$

where $\bar{F}(k)$ is the Fourier transform of $f'(x)$, which is defined to be zero for $x < 0$, $x > l$. Hence, show that

$$\frac{d\bar{\phi}}{dy} = U\bar{F}(k)e^{-\beta|k|y}.$$

From Exercise 4.4, the Fourier transform of $\beta y/(x^2 + \beta^2 y^2)$ is $\pi e^{-\beta|k|y}$; use this result to determine $\partial\phi/\partial y$ and hence show that

$$\phi(x,y) = \frac{U}{2\pi\beta} \int_{-L}^{L} f'(\xi) \log((x-\xi)^2 + \beta^2 y^2) \, d\xi.$$

*4.18 (i) Solve the problem of subsonic flow past a slender axisymmetric body at zero incidence by using Fourier transforms as follows. If $\bar{\phi}$ is defined as

$$\bar{\phi}(r,k) = \int_{-\infty}^{\infty} \phi(x,r)e^{ikx} \, dx,$$

where the velocity potential of the flow is $Ux + \varepsilon\phi$, show that $\bar{\phi}$ satisfies the equation

$$\frac{d^2\bar{\phi}}{dr^2} + \frac{1}{r}\frac{d\bar{\phi}}{dr} - \beta^2 k^2 \bar{\phi} = 0.$$

The solution of this equation which tends to zero as $r \to \infty$ is the Bessel function $K_0(\beta|k|r)$. By writing

$$\bar{\phi} = \bar{A}(k)K_0(\beta|k|r)$$

and using the fact that $K_0(\beta|k|r) \sim -\log r$ as $r \to 0$ in the boundary condition (4.57), show that $\bar{A}(k)$ is the Fourier transform of $A(x)$, where

$$A(x) = -\varepsilon U R(x)R'(x) \quad \text{for } 0 < x < l.$$

Given that $K_0(\beta|k|r)$ is the Fourier transform of $1/2\sqrt{x^2 + \beta^2 r^2}$, show that

$$\phi(x,r) = \int_0^l \frac{-\varepsilon U R(\xi)R'(\xi)}{2((x-\xi)^2 + \beta^2 r^2)^{1/2}} \, d\xi.$$

(ii) A slender body $r = \varepsilon R(x)$ is now placed in a subsonic stream which makes a small angle α to the axis of the body. If the free-stream velocity is

$$(U\alpha\cos\theta, -U\alpha\sin\theta, U)$$

in the (r, θ, x) directions in cylindrical polar coordinates, show that the boundary condition on the body is

$$\frac{\partial\phi}{\partial r} = UR'(x) - \frac{U\alpha}{\varepsilon}\cos\theta. \qquad (*)$$

Assuming that α/ε is of $O(1)$ and using the Fourier transform, as in part (i), show that

$$\bar{\phi} = \bar{A}(k)K_0(\beta|k|r) + \bar{B}(k)\frac{\partial}{\partial r}(K_0(\beta|k|r))\cos\theta.$$

If $B(x)$ is the inverse transform of $\bar{B}(k)$, show that

$$B(x) = -U\alpha\varepsilon(R(x))^2,$$

and deduce that the full solution is

$$\phi(r,\theta,x) = -\frac{\varepsilon U}{2}\int_0^l \frac{R(\xi)R'(\xi)\,d\xi}{((x-\xi)^2 + \beta^2 r^2)^{1/2}}$$
$$-\frac{U\alpha\varepsilon\cos\theta}{2}\frac{\partial}{\partial r}\int_0^l \frac{(R(\xi))^2\,d\xi}{((x-\xi)^2 + \beta^2 r^2)^{1/2}}.$$

R4.19 A thin wing is placed at a small angle of incidence α in a steady supersonic stream, so that the wing is given by

$$y = \varepsilon f_\pm(x) - \alpha x \quad \text{for } 0 < x < l,$$

where $\alpha = O(\varepsilon)$. Show that the drag is

$$\frac{\rho_0 U^2}{B}\int_0^l [(\varepsilon f_+'(x) - \alpha)^2 + (\varepsilon f_-'(x) - \alpha)^2]\,dx.$$

Hence, confirm that the drag on a flat plate of length l at a small angle of incidence α is $(2\rho_0 U^2/B)\alpha^2 l$.

Show that if $\alpha = 0$ and the wing has a cross section

$$f_+ = -f_- = \begin{cases} mx, & 0 < x < \dfrac{h}{m} \\ \dfrac{h(l-x)}{l-h/m}, & \dfrac{h}{m} < x < l \end{cases}$$

with a given thickness h, then the drag is minimized for a diamond shape with $h/m = l/2$.

4.20 A plane high-frequency wave, given by $\Phi = e^{-ikx}$, is incident from the right on a parabolic reflector $y^2 = 4x$. If the reflected wave is given by $\Phi = e^{iku}$, show that

$$x = \tan^2\left(\frac{s}{2}\right), \qquad y = 2\tan\left(\frac{s}{2}\right), \qquad u = -\tan^2\left(\frac{s}{2}\right), \qquad (-\pi < s < \pi)$$

is suitable boundary data for u. By showing that $p = \cos s$ and $q = -\sin s$ and solving Charpit's equations (4.67), show that the reflected rays are given by

$$x = 2\tau\cos s + \tan^2\left(\frac{s}{2}\right), \qquad y = -2\tau\sin s + 2\tan\left(\frac{s}{2}\right), \qquad u = 2\tau - \tan^2\left(\frac{s}{2}\right).$$

Deduce that all the reflected rays pass through the focus $(1,0)$.

4.21 Suppose that in (4.71), the function f is spherically symmetric so that $f(x, y, z) = F(r)$, where $x^2 + y^2 + z^2 = r^2$. Show that at any point $(0, 0, z)$,

$$\phi(0, 0, z, t) = \frac{t}{4\pi} \int_0^{2\pi} \int_0^\pi F((z^2 + 2zct \cos\theta + c^2 t^2)^{1/2}) \sin\theta \, d\theta \, d\phi$$

$$= \frac{1}{2cz} [g(|z + ct|) - g(|z - ct|)],$$

where $g'(r) = rF(r)$. By rotating the axes, deduce that

$$\phi(x, y, z, t) = \frac{1}{2cr} [g(r + ct) - g(|r - ct|)].$$

Verify that this gives a spherically symmetric solution of the wave equation which satisfies $\phi = 0$ and $\partial\phi/\partial t = F(r)$ at $t = 0$.

4.22 Suppose that, in (4.71), the function $f(x, y, z)$ is independent of z. Write $\rho = ct \sin\theta$ when $|\rho| < ct$ and show that

$$\phi(x, y, t) = \frac{1}{2\pi c} \int_0^{2\pi} \int_0^{ct} f(x + \rho\cos\phi, y + \rho\sin\phi) \frac{\rho \, d\rho \, d\phi}{\sqrt{c^2 t^2 - \rho^2}}$$

$$= \frac{1}{2\pi c} \int\int_S \frac{f(\xi, \eta) \, d\xi \, d\eta}{(c^2 t^2 - (x - \xi)^2 - (y - \eta)^2)^{1/2}},$$

where S is the interior of the circle $(x - \xi)^2 + (y - \eta)^2 = c^2 t^2$. Hence, show that there is no sharp termination of the wave in the case when $f(\xi, \eta)$ vanishes outside some bounded region in the (ξ, η) plane.

Show also that if $f(\xi, \eta)$ is localized near $\xi = 0$, $\eta = 0$, we retrieve the solution (4.70), where, for some constant λ,

$$\phi = \begin{cases} \dfrac{\lambda}{(c^2 t^2 - r^2)^{1/2}}, & r < ct \\ 0, & r > ct, \end{cases}$$

where $r^2 = x^2 + y^2$.

*4.23 (i) The spherically symmetric wave equation in n dimensions is

$$\frac{\partial^2\phi}{\partial r^2} + \frac{n - 1}{r}\frac{\partial\phi}{\partial r} = \frac{1}{c^2}\frac{\partial^2\phi}{\partial t^2}.$$

Show that when n is odd, the general solution is

$$\left(\frac{1}{r}\frac{\partial}{\partial r}\right)^{\frac{n-3}{2}} \frac{f(ct \pm r)}{r}.$$

(ii) From (4.61), show that the general solution for outward propagating axisymmetric waves in two dimensions is

$$\phi = \int_r^{ct} \frac{f(ct - s) \, ds}{\sqrt{s^2 - r^2}},$$

where $m(z/c) = f(z)$.

It can be shown that in an even number of dimensions n, the corresponding general solution is

$$\left(\frac{1}{r}\frac{\partial}{\partial r}\right)^{(n-2)/2}\left[\int_r^{ct}\frac{f(ct-s)\,ds}{\sqrt{s^2-r^2}}\right],$$

which is the same as the result of (1) in terms of pseudo-differential operators.

4.24 By taking the integral in (4.58) from $x - ir$ to $x + ir$ and assuming that $h(\xi)$ is analytic throughout, show that

$$\phi(x,r) = \int_{x-i\beta r}^{x+i\beta r}\frac{h(\xi)\,d\xi}{((x-\xi)^2+\beta^2r^2)^{1/2}}$$

$$= \int_0^\pi ih(x+i\beta r\cos\theta)\,d\theta$$

satisfies the axisymmetric equation (4.56). Verify that ϕ is analytic on $r = 0$ and that

$$\phi(x,r) = \frac{1}{\pi}\int_0^\pi \phi(x+i\beta r\cos\theta,0)\,d\theta. \qquad (\dagger)$$

Note the difference between this result, which holds for axisymmetric potentials that are analytic at $r = 0$ for all x, and the representation (4.58), where $\phi \sim -2h(x)\log r$ as $r \to 0$.

4.25 Show that

$$\phi(x,r) = \int_0^{x-Br}\frac{m(\xi)\,d\xi}{\sqrt{(x-\xi)^2-B^2r^2}} = \int_0^{\cosh^{-1}(x/Br)}m(x-Br\cosh t)\,dt$$

and hence confirm that

$$\frac{\partial^2\phi}{\partial r^2} + \frac{1}{r}\frac{\partial\phi}{\partial r} - B^2\frac{\partial^2\phi}{\partial x^2} = 0.$$

5

Nonlinear Waves in Fluids

5.1 Introduction

We have already encountered several deficiencies in the theories presented in Chapter 4 that indicate the limitations of the linear approximation. In particular, we have seen that the linear theory cannot deal with the following situations:

(i) The transition from subsonic to supersonic flow past a thin wing.
(ii) The strange behavior in a converging–diverging nozzle when sonic conditions are attained.
(iii) The response of a system near resonance.

More generally, we may also ask what happens when bodies which are thick or "blunt-nosed" are placed in a compressible stream, or when the elevation of a surface gravity wave is comparable to the depth of the water, or when the amplitude of the motion of a gas in a resonator is comparable to the length scale of the system.

In this chapter, we will consider three specific nonlinear models: unsteady one-dimensional gasdynamics, two-dimensional steady gasdynamics, and shallow water theory. However, before we embark on these models, we will discuss a simple paradigm example of a nonlinear system that will help us to understand the more complicated systems that will follow.

We consider the equation

$$\frac{\partial u}{\partial t} + u \frac{\partial u}{\partial x} = 0, \qquad (5.1)$$

defined for $-\infty < x < \infty$ and $t > 0$, and with $u(x,0) = u_0(x)$ prescribed.[1] This nonlinear equation can be solved exactly. The general solution is $u = F(x - ut)$ for any function F and so the solution

$$u = u_0(x - ut) \tag{5.2}$$

describes the solution implicitly. An alternative approach is to note that, along the characteristics of (5.1),

$$\frac{dx}{dt} = u, \qquad \frac{du}{dt} = 0,$$

and so u is constant on a characteristic, which is therefore a straight line. Thus, the characteristics can be drawn just by using the initial slopes, as given by $u_0(x)$, and an example is shown in Figure 5.1.

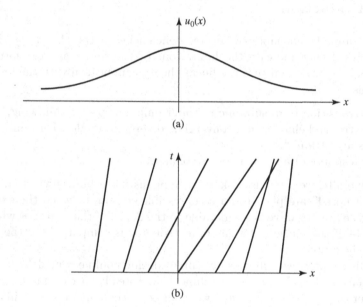

(a)

(b)

Fig. 5.1. (a) Initial data for equation (5.1). (b) Characteristics of equation (5.1).

We are immediately confronted by one of the fundamental difficulties of nonlinear hyperbolic partial differential equations, which is that since the slope of the characteristic depends on the solution, it is possible for characteristics to *intersect*, leading to a multivalued solution for u.

---

[1] This is an example of what is often known as the *kinematic wave equation*, namely

$$\frac{\partial u}{\partial t} + \frac{\partial}{\partial x}(f(u)) = 0$$

for some function f.

If $u_0(x)$ is smooth and has finite slope everywhere, then the solution (5.2) will hold for small values of t, but, except in rare situations, such as when u_0 is monotonic increasing for all x, u will eventually become multivalued as a result of intersecting characteristics. Of course, it may be possible to allow multivalued solutions if u describes the profile of, say, a water wave, but, in general, multivalued solutions are not physically acceptable; the remedy for this situation is the introduction of a discontinuity or shock wave as will be described in Chapter 6.

5.2 Models for Nonlinear Waves

5.2.1 One-dimensional Unsteady Gasdynamics

We now consider some exact solutions of the full equations for the flow of a perfect compressible gas. The first case we consider is that of one-dimensional unsteady flow, and in this case, (2.6), (2.7), and (2.18) reduce to

$$\frac{\partial \rho}{\partial t} + \frac{\partial}{\partial x}(\rho u) = 0, \tag{5.3}$$

$$\frac{\partial u}{\partial t} + u\frac{\partial u}{\partial x} + \frac{1}{\rho}\frac{\partial p}{\partial x} = 0 \tag{5.4}$$

and

$$\frac{d}{dt}\left(\frac{p}{\rho^\gamma}\right) = 0. \tag{5.5}$$

We assume that the flow is homentropic, as it would be if the gas was initially in a uniform state. This assumption means that, from (5.5), p/ρ^γ is constant, and since $c^2 = dp/d\rho$, we can deduce that

$$\frac{dp}{p} = \gamma\frac{d\rho}{\rho} = \frac{2\gamma\,dc}{(\gamma-1)c}. \tag{5.6}$$

Using (5.6), (5.3) and (5.4) can be written in terms of u and c alone as

$$\frac{2}{\gamma-1}\frac{\partial c}{\partial t} + \frac{2u}{\gamma-1}\frac{\partial c}{\partial x} + c\frac{\partial u}{\partial x} = 0 \tag{5.7}$$

and

$$\frac{\partial u}{\partial t} + u\frac{\partial u}{\partial x} + \frac{2c}{\gamma-1}\frac{\partial c}{\partial x} = 0. \tag{5.8}$$

Adding and subtracting (5.7) and (5.8) leads to

$$\left(\frac{\partial}{\partial t} + (u \pm c)\frac{\partial}{\partial x}\right)\left(u \pm \frac{2c}{\gamma-1}\right) = 0; \tag{5.9}$$

hence, we can see that

$$u \pm \frac{2c}{\gamma - 1} \quad \text{is constant on} \quad \frac{dx}{dt} = u \pm c. \qquad (5.10)$$

The quantities $u \pm 2c/(\gamma-1)$ are called the *Riemann invariants*, and the curves $dx/dt = u \pm c$ are the *characteristics*[2] of the second-order hyperbolic system of equations (5.9). More generally, for non-homentropic flow, the three equations (5.3)–(5.5) form a *third-order hyperbolic system*, to be defined precisely in Section 5.2.3, with characteristics given by $dx/dt = u \pm c$ and $dx/dt = u$, where $c^2 = \gamma p/\rho$. The "third" characteristic is the particle path.

5.2.2 Two-dimensional Steady Homentropic Gasdynamics

Our second example is two-dimensional steady compressible flow. We again assume that the flow is homentropic so that p/ρ^γ is constant and then (2.6) and (2.7) can be written as

$$\frac{\partial}{\partial x}(\rho u) + \frac{\partial}{\partial y}(\rho v) = 0, \qquad (5.11)$$

$$u\frac{\partial u}{\partial x} + v\frac{\partial u}{\partial y} + \frac{c^2}{\rho}\frac{\partial \rho}{\partial x} = 0 \qquad (5.12)$$

and

$$u\frac{\partial v}{\partial x} + v\frac{\partial v}{\partial y} + \frac{c^2}{\rho}\frac{\partial \rho}{\partial y} = 0. \qquad (5.13)$$

We use the same ideas to simplify these equations as we did in the unsteady one-dimensional case, but because there are three equations now rather than two, the technical details are more complicated. Nevertheless, we can easily eliminate $\partial \rho/\partial x$ and $\partial \rho/\partial y$ from (5.11)–(5.13) to obtain

$$(c^2 - u^2)\frac{\partial u}{\partial x} - uv\left(\frac{\partial u}{\partial y} + \frac{\partial v}{\partial x}\right) + (c^2 - v^2)\frac{\partial v}{\partial y} = 0. \qquad (5.14)$$

To get a closed system for u and v, we can note that the flow will be irrotational by Crocco's theorem (Exercise 2.5); hence, c^2 is given from Bernoulli's equation (2.25) as

$$c^2 = c_0^2 - \frac{\gamma - 1}{2}(u^2 + v^2), \qquad (5.15)$$

where c_0 is the value of c when the flow is brought to rest homentropically, and

$$\frac{\partial u}{\partial y} - \frac{\partial v}{\partial x} = 0. \qquad (5.16)$$

---

[2] For brevity, we will sometimes refer to these two families of characteristics as the *positive* and *negative* characteristic, respectively.

We could now define a potential function ϕ and then (5.14) and (5.15) yield

$$\left(c^2 - \left(\frac{\partial \phi}{\partial x}\right)^2\right)\frac{\partial^2 \phi}{\partial x^2} - 2\frac{\partial \phi}{\partial x}\frac{\partial \phi}{\partial y}\frac{\partial^2 \phi}{\partial x \partial y} + \left(c^2 - \left(\frac{\partial \phi}{\partial y}\right)^2\right)\frac{\partial^2 \phi}{\partial y^2} = 0, \quad (5.17)$$

where $c^2 = c_0^2 - [(\gamma - 1)/2](\nabla \phi)^2$. Equation (5.17) is a second-order quasi-linear equation for ϕ to which we will return later, but, for now, we work with the second-order system (5.14) and (5.16) and use the methods of Section 5.2.1 to try to write these equations in a more convenient form. Adding (5.14) and a multiple of (5.16), we find that by choosing the multipliers to be $\pm c\sqrt{u^2 + v^2 - c^2}$, the equations emerge in a form that can be integrated (Exercise 5.2); however, to proceed, we do have to make the all-important assumption that $u^2 + v^2 > c^2$. We find that

$$(c^2 - u^2)\,du - (uv \pm c\sqrt{u^2 + v^2 - c^2})\,dv = 0 \quad (5.18)$$

on the characteristic curves given by

$$\frac{dy}{dx} = \frac{-uv \pm c\sqrt{u^2 + v^2 - c^2}}{c^2 - u^2}. \quad (5.19)$$

If we introduce the new variables

$$\mu = \sin^{-1}\left(\frac{c}{\sqrt{u^2 + v^2}}\right) \quad \text{and} \quad \theta = \tan^{-1}\left(\frac{v}{u}\right),$$

we find that (5.18) can be integrated to give the Riemann invariants

$$\theta \pm \left[\mu + \frac{1}{\lambda}\tan^{-1}(\lambda \cot \mu)\right], \quad (5.20)$$

where $\lambda^2 = (\gamma - 1)/(\gamma + 1)$; these invariants are constant on the characteristics given by

$$\frac{dy}{dx} = \tan(\theta \mp \mu), \quad (5.21)$$

respectively. The angle μ defined above is called the *Mach angle*, and from (5.21), we see that the characteristics always make an angle $\mp\mu$ with the streamlines, and these will again be called the negative and positive characteristics, respectively. Note that, as can be seen from (5.18) and (5.19), these characteristics and Riemann invariants are real in view of our assumption that $u^2 + v^2 \geq c^2$, so that the flow is supersonic throughout. Equations (5.11)–(5.13) are a hyperbolic system as long as $u^2 + v^2 > c^2$, and the characteristics will be given by (5.19) and $dy/dx = v/u$. The third characteristic is, naturally, the streamline.

5.2.3 Shallow Water Theory

It would be nice if we could treat nonlinear surface gravity waves as described by (3.8), (3.10) and (3.11) similarly, but this system is even more difficult to analyze than the models in Sections 5.2.1 and 5.2.2. The only way we can make headway here is if we restrict ourselves to analyzing the effect of nonlinearity on water that is *shallow*. What we mean by this is that the mean depth of the water, h, is comparable to the amplitude of the waves but small compared to their wavelength λ.

To see the implications of these assumptions, we non-dimensionalize the variables by writing

$$x = \lambda X, \qquad z = hZ, \qquad u = U\hat{u}, \qquad t = \frac{\lambda}{U}T,$$

where U is a typical horizontal velocity; for the moment, we are only considering two-dimensional flows with z measured along the upward vertical. From the continuity equation for an incompressible fluid (2.6), the appropriate non-dimensionalization for the vertical component of velocity is

$$w = \left(\frac{hU}{\lambda}\right)\hat{w},$$

and from the x component of the momentum equation (2.7), the appropriate scaling for the pressure is

$$p = p_0 + \rho U^2 \hat{p},$$

where p_0 is the pressure in the atmosphere. Now, (2.6) and (2.7) become

$$\frac{\partial \hat{u}}{\partial X} + \frac{\partial \hat{w}}{\partial Z} = 0, \tag{5.22}$$

$$\frac{\partial \hat{u}}{\partial T} + \hat{u}\frac{\partial \hat{u}}{\partial X} + \hat{w}\frac{\partial \hat{u}}{\partial Z} = -\frac{\partial \hat{p}}{\partial X} \tag{5.23}$$

and

$$\frac{\partial \hat{p}}{\partial Z} + \frac{gh}{U^2} = O\left(\frac{h^2}{\lambda^2}\right). \tag{5.24}$$

From the last equation, we see that if $h \ll \lambda$, the fluid inertia terms in the Z direction can be neglected and also that the appropriate choice for U is $\sqrt{gh}$. Hence, the pressure is *hydrostatic* to lowest order. Thus, reverting to dimensional variables and integrating, we have

$$p = -\rho g z + \rho g \eta + p_0, \tag{5.25}$$

where $z = \eta(x,t)$ is the equation of the surface of the water. Hence, substituting for p in (5.23) and writing that equation in dimensional variables, we obtain

$$\frac{du}{dt} = -g\frac{\partial \eta}{\partial x}, \tag{5.26}$$

showing that the convective derivative of u is independent of z. Thus, if u is initially independent of z, it will remain independent of z for all time.[3] Then, writing $u = u(x,t)$, (5.26) reduces to

$$\frac{\partial u}{\partial t} + u\frac{\partial u}{\partial x} + g\frac{\partial \eta}{\partial x} = 0. \tag{5.27}$$

Now, we can also integrate (5.22) with respect to z and this leads to

$$w = -\frac{\partial u}{\partial x}z \tag{5.28}$$

if we assume a flat bottom with $w = 0$ on $y = 0$. Now, we finally need to apply the kinematic boundary condition (3.10) at the surface $z = \eta$ to get

$$w = \frac{\partial \eta}{\partial t} + u\frac{\partial \eta}{\partial x}, \tag{5.29}$$

and (5.28) and (5.29) lead to

$$\frac{\partial \eta}{\partial t} + \frac{\partial}{\partial x}(u\eta) = 0. \tag{5.30}$$

Equations (5.27) and (5.30) can be thought of as statements of conservation of momentum and mass, respectively, averaged through the depth of the water; they are two nonlinear equations for $u(x,t)$ and $\eta(x,t)$. It is convenient to write $s^2 = g\eta$ and then add and subtract these two equations to obtain

$$\left(\frac{\partial}{\partial t} + (u \pm s)\frac{\partial}{\partial x}\right)(u \pm 2s) = 0. \tag{5.31}$$

We have now arrived at a formulation of the problem which is very like that obtained in Section 5.2.1 and shows that

$$u \pm 2s = \text{ constant}$$

on the characteristics

$$\frac{dx}{dt} = u \pm s.$$

In fact, the model (5.31) is *identical* with the one-dimensional unsteady gas-dynamic equations (5.9) if we put $c = s$ and $\gamma = 2$ (remember that $\gamma = 1.4$ for air).

Reviewing the three sets of equations (5.3)–(5.5), (5.11)–(5.13), and (5.27), (5.30), we can look at a more general methodology for dealing with these systems. Each of these sets of equations can be written in the form

$$A\frac{\partial \mathbf{u}}{\partial X} + B\frac{\partial \mathbf{u}}{\partial Y} = 0, \tag{5.32}$$

---

[3] It is interesting to relate this assumption to that of irrotationality (Exercises 5.4 and 5.7).

where $\mathbf{u}$ is an n-vector and A and B are $n \times n$ matrices whose entries are functions of the components of $\mathbf{u}(X,Y)$. We note that if A is non-singular and $A^{-1}B$ has an eigenvalue λ and a left eigenvector $\mathbf{l}$, so that

$$\det(\lambda A - B) = 0 \quad \text{and} \quad \mathbf{l}A^{-1}B = \lambda \mathbf{l},$$

then

$$\mathbf{l} \cdot \frac{\partial \mathbf{u}}{\partial X} + \lambda \mathbf{l} \cdot \frac{\partial \mathbf{u}}{\partial Y} = 0.$$

Thus, $\int \mathbf{l} \cdot d\mathbf{u}$ is constant on the curve $dY/dX = \lambda$, and this is exactly equivalent to the procedure that we have used already for each of the above problems. Indeed, this is the starting point for the theory of hyperbolic systems; (5.32) is called hyperbolic if all n eigenvalues of $A^{-1}B$ are real and distinct, $dY/dX = \lambda$ are the characteristic curves, and $\int \mathbf{l} \cdot d\mathbf{u}$ are the Riemann invariants (Ockendon et al. [9]).

Finally, we remark that if we linearize the shallow water equations (5.27) and (5.30) by assuming that u and $\eta - h$ are small,[4] we are led to the familiar one-dimensional wave equation

$$\frac{\partial^2 u}{\partial t^2} = gh \frac{\partial^2 u}{\partial x^2}. \tag{5.33}$$

Solutions of (5.33) are called two-dimensional *tidal waves* and it is easy to verify that in most seas and oceans, the tides move along the shore at a speed of $O(\sqrt{gh})$. It has already been observed in Section 4.4.1 of Chapter 4 that as $h \to 0$ in the Stokes wave solution (4.25), the dispersion relation reduces to

$$\omega^2 = gh|k|^2,$$

exactly as predicted from (5.33). Thus, long, small-amplitude waves on shallow water are non-dispersive and the phase and group velocities for such waves are both $\sqrt{gh}$. In fact, most tidal waves propagate in two dimensions and in this case (5.33) is replaced by

$$\frac{\partial^2 \eta}{\partial t^2} = gh \left(\frac{\partial^2 \eta}{\partial x^2} + \frac{\partial^2 \eta}{\partial y^2} \right),$$

which is also the equation for waves on a membrane.

5.2.4 *Nonlinearity and Dispersion

5.2.4.1 The Korteweg–de Vries Equation

The full model for nonlinear surface waves, described by (3.8), (3.10), and (3.11), can be simplified in other parameter regimes that are distinct from

---

[4] We can be more precise about how small these quantities are; if we define $(\eta - h)/h = O(\varepsilon)$, then u will be $O(\varepsilon\sqrt{gh})$.

Stokes waves, shallow water theory, or tidal theory, and a fascinating situation occurs if we consider the *long-time evolution* of a tidal wave. To illustrate why the solution obtained in the last section may not be valid over long times, we first consider the model equation

$$\frac{\partial \phi}{\partial t} + \frac{\partial \phi}{\partial x} = \varepsilon \frac{\partial^2 \phi}{\partial x^2}, \tag{5.34}$$

where ε is a small parameter. If we write $\phi \sim \phi_0 + \varepsilon \phi_1 + \cdots$ and expand in powers of ε, we find that

$$\phi_0 = f(x - t)$$

and

$$\frac{\partial \phi_1}{\partial t} + \frac{\partial \phi_1}{\partial x} = f''(x - t),$$

so that

$$\phi_1 = t f''(x - t) + g(x - t),$$

where f and g are arbitrary functions. Thus, we see that when $t \sim O(\varepsilon^{-1})$, $\varepsilon \phi_1$ will be of the same order as ϕ_0 and the expansion for small ε is no longer valid. To find a solution valid for such timescales, we need to rewrite (5.34) in terms of new independent variables $\xi = x - t$ and $\tau = \varepsilon t$ so that it becomes

$$\frac{\partial \phi}{\partial \tau} = \frac{\partial^2 \phi}{\partial \xi^2}. \tag{5.35}$$

Now, we need to solve (5.35) with the initial condition $\phi = f(\xi)$ at $\tau = 0$ to get a solution to (5.34) which is valid for all times. It is interesting to note that for this linear model, we can use Fourier transforms to obtain the long-time behavior and thereby assess the usefulness of (5.35) (Exercise 5.12).

We now show how the same method can be applied to tidal waves. We recall that Stokes waves were derived under the assumption that the ratio of amplitude to depth, $a/h = \varepsilon$, is small. Shallow water theory assumes that the ratio of depth to wavelength, $h/\lambda = \delta$, is small, and tidal wave theory assumes that both ε and δ are small. Using the same scalings as for tidal wave theory in Section 5.2.3, we non-dimensionalize x with λ, z with h, η with a, t with $\lambda/\sqrt{gh}$, and u with $\varepsilon\sqrt{gh}$. Then, the appropriate scaling for ϕ is $\varepsilon\lambda\sqrt{gh}$ and (3.8), (3.10) and (3.11) become, in non-dimensional form,

$$\frac{\partial^2 \phi}{\partial z^2} + \delta^2 \frac{\partial^2 \phi}{\partial x^2} = 0, \tag{5.36}$$

with

$$\frac{\partial \phi}{\partial t} + \eta + \frac{\varepsilon}{2\delta^2}\left(\left(\frac{\partial \phi}{\partial z}\right)^2 + \delta^2 \left(\frac{\partial \phi}{\partial x}\right)^2\right) = 0 \tag{5.37}$$

and

$$\frac{\partial \phi}{\partial z} = \delta^2 \frac{\partial \eta}{\partial t} + \varepsilon\delta^2 \frac{\partial \phi}{\partial x} \cdot \frac{\partial \eta}{\partial x} \tag{5.38}$$

on $z = \varepsilon\eta$. The final boundary condition on the bottom is

$$\frac{\partial \phi}{\partial z} = 0 \qquad (5.39)$$

on $z = -1$. We can see at once that we can retrieve Stokes waves by taking $\delta = 1$ and $\varepsilon = 0$.

When both ε and δ are small, we find that in order to satisfy (5.36) and (5.39), we must write

$$\phi(x, z, t; \varepsilon, \delta) \sim \phi_0(x, t; \varepsilon) + \delta^2 \left(A(x, t; \varepsilon) - \left(\frac{1}{2} z^2 + z \right) \frac{\partial^2 \phi_0}{\partial x^2} \right) + \cdots, \quad (5.40)$$

for some function A. Then, writing $\eta \sim \eta_0(x, t; \varepsilon) + \delta^2 \eta_1(x, t; \varepsilon) + \cdots$, (5.37) and (5.38) lead to

$$\frac{\partial \phi_0}{\partial t} + \eta_0 = O(\varepsilon, \delta^2)$$

and

$$\frac{\partial^2 \phi_0}{\partial x^2} + \frac{\partial \eta_0}{\partial t} = O(\varepsilon, \delta^2),$$

and, hence, when $\varepsilon \ll 1$, to the tidal wave equation (5.33). When $\varepsilon = 1$ and $\delta \to 0$ we can derive the equations for shallow water, (5.27) and (5.30), in a similar way (Exercise 5.7). However, the example (5.34) leads us to examine whether the tidal wave approximation will really be valid for large times. We see that as t increases, the time derivative in (5.37) will become as small as some of the neglected terms, and to deal with this, we have to take into account even more terms in the expansion (5.40) for ϕ.

Again, using (5.36) and (5.39) but going to the next term in the expansion in δ^2 gives

$$\phi \sim \phi_0(x, t; \varepsilon) + \delta^2 \left(A(x, t; \varepsilon) - \left(\frac{1}{2} z^2 + z \right) \frac{\partial^2 \phi_0}{\partial x^2} \right)$$
$$+ \delta^4 \left(B(x, t; \varepsilon) - \left(\frac{1}{2} z^2 + z \right) \frac{\partial^2 A}{\partial x^2} + \left(\frac{1}{24} z^4 + \frac{1}{6} z^3 - \frac{1}{3} z \right) \frac{\partial^4 \phi_0}{\partial x^4} \right) + \cdots.$$
$$(5.41)$$

Then, remembering that the surface condition is applied on $z = \varepsilon\eta$, (5.37) is

$$\frac{\partial \phi_0}{\partial t} + \eta_0 + \frac{1}{2}\varepsilon \left(\frac{\partial \phi_0}{\partial x} \right)^2 + \delta^2 \left(\frac{\partial A}{\partial t} + \eta_1 \right) = O(\varepsilon\delta^2), \qquad (5.42)$$

and (5.38) gives

$$\frac{\partial \eta_0}{\partial t} + \frac{\partial^2 \phi_0}{\partial x^2} + \varepsilon \left(\eta_0 \frac{\partial^2 \phi_0}{\partial x^2} + \frac{\partial \phi_0}{\partial x} \frac{\partial \eta_0}{\partial x} \right)$$
$$+ \delta^2 \left(\frac{\partial^2 A}{\partial x^2} + \frac{1}{3} \frac{\partial^4 \phi_0}{\partial x^4} + \frac{\partial \eta_1}{\partial t} \right) + O(\varepsilon\delta^2, \delta^4) = 0. \qquad (5.43)$$

At this point, we need to rescale $t = \varepsilon^{-1}\tau$, and remembering that the tidal wave generates solutions for η_0 and ϕ_0 of the form $f(x-t)+g(x+t)$, we follow just the right-traveling wave by writing $x - t = \xi$ and keeping ξ of $O(1)$. We also have to decide on the relation between our two small parameters ε and δ. It is clear from (5.42) and (5.43) that the most interesting case will be when $\varepsilon = O(\delta^2)$ and accordingly we write $\delta^2 = \kappa\varepsilon$. We thus write (5.42) and (5.43) as

$$-\frac{\partial\phi_0}{\partial\xi} + \eta_0 + \varepsilon\left(\frac{1}{2}\left(\frac{\partial\phi_0}{\partial\xi}\right)^2 - \kappa\frac{\partial A}{\partial\xi} + \kappa\eta_1 + \frac{\partial\phi_0}{\partial\tau}\right) = O(\varepsilon^2)$$

and

$$-\frac{\partial\eta_0}{\partial\xi} + \frac{\partial^2\phi_0}{\partial\xi^2} + \varepsilon\left(\eta_0\frac{\partial^2\phi_0}{\partial\xi^2} + \frac{\partial\phi_0}{\partial\xi}\frac{\partial\eta_0}{\partial\xi}\right.$$
$$\left. + \kappa\frac{\partial^2 A}{\partial\xi^2} + \frac{\kappa}{3}\frac{\partial^4\phi_0}{\partial\xi^4} - \kappa\frac{\partial\eta_1}{\partial\xi} + \frac{\partial\eta_0}{\partial\tau}\right) = O(\varepsilon^2).$$

Hence, $\eta_0 = \partial\phi_0/\partial\xi$, and by adding the ξ derivative of the first equation to the second, we finally obtain that

$$\frac{\partial\eta_0}{\partial\tau} + \frac{3}{2}\eta_0\frac{\partial\eta_0}{\partial\xi} + \frac{\kappa}{6}\frac{\partial^3\eta_0}{\partial\xi^3} = 0 \qquad (5.44)$$

is the condition for (5.42) and (5.43) to be simultaneously valid.[5] This is the *Korteweg–de Vries (KdV) equation* and it is valid for times of $O(\lambda^3/h^3)\sqrt{h/g}$ if we assume that $\frac{a}{h} = O\left((\frac{h}{\lambda})^2\right)$. The equation represents a balance between the linear "tidal wave" term $\partial\eta_0/\partial\tau$, the nonlinear "shallow water" term $\eta_0(\partial\eta_0/\partial\xi)$, and dispersive "Stokes wave" term $\partial^3\eta_0/\partial\xi^3$. The KdV equation has *traveling wave* solutions which can be obtained by writing $\eta_0 = f(\xi - c\tau)$. Substituting into (5.44) and integrating once gives

$$\frac{\kappa}{6}\frac{d^2 f}{d\chi^2} = cf - \frac{3}{4}f^2 + k_1, \qquad (5.45)$$

where k_1 is a constant and $\chi = \xi - c\tau = x - (1 + c\varepsilon)t$. If we require that, as $\chi \to \pm\infty$, the solution f and its derivatives tend to zero, then $k_1 = 0$ and (5.45) can be integrated to give

$$f = 2c\,\mathrm{sech}^2\left(\sqrt{\frac{3c}{2\kappa}}(\chi + d)\right), \qquad (5.46)$$

where d is a constant. Because of its behavior as $|\chi| \to \infty$, this profile is called a *solitary wave*. It is a traveling wave of constant speed and shape which can easily be observed in either a long straight canal[6] or in a laboratory. Even

---

[5] The step leading to (5.44) is yet another example of the Fredholm alternative.

[6] Scott Russell famously observed "a great wave of elevation" while riding his horse along the towpath of a canal in 1845 ([18]).

more remarkable is the *theory of solitons*, which was opened up by the study of the KdV equation. A soliton is a solitary wave that has certain very special properties in addition to its permanent shape, and (5.46) was the first solitary wave that was discovered to have these special properties. We will return to this topic briefly at the end of this section after we have considered another important example of gradual nonlinear modulation.

We remark that although nonlinearity destroys the predictions of linear tidal theory for long times, it does not destroy most of the predictions of linear Stokes wave theory on deeper water. Indeed, as long as we use the stationary phase arguments of Chapter 4 correctly, Stokes wave theory is *uniformly* valid for all times, with just one exception which we will now discuss.

5.2.4.2 The Nonlinear Schrödinger Equation

Another situation in which the linear approximation breaks down for long times is the *periodic* solution for Stokes waves on deep water. Once again, we illustrate the way this happens by considering a simple model equation—in this case, an ordinary differential equation with a periodic solution in the linear approximation and a seemingly small quadratic nonlinearity. Thus, we consider the equation

$$\frac{d^2x}{dt^2} + x = \varepsilon x^2, \tag{5.47}$$

and put $x \sim x_0 + \varepsilon x_1 + \varepsilon^2 x_2 + \cdots$ to obtain the equations

$$\frac{d^2x_0}{dt^2} + x_0 = 0, \tag{5.48}$$

$$\frac{d^2x_1}{dt^2} + x_1 = x_0^2 \tag{5.49}$$

and

$$\frac{d^2x_2}{dt^2} + x_2 = 2x_0x_1. \tag{5.50}$$

Solving (5.48) and (5.49) gives

$$x_0 = Ae^{it} + A^*e^{-it} \tag{5.51}$$

and

$$x_1 = Be^{it} + B^*e^{-it} - \frac{1}{3}A^2e^{2it} + 2AA^* - \frac{1}{3}A^{*2}e^{-2it}, \tag{5.52}$$

where A and B are complex constants and $*$ denotes complex conjugate.[7] Now we find that equation (5.50) becomes

$$\frac{d^2x_2}{dt^2} + x_2 = \frac{10}{3}A^2A^*e^{it} + \frac{10}{3}AA^{*2}e^{-it} + \text{constant} + \text{terms in } e^{\pm 2it} \text{ and } e^{\pm 3it}.$$

[7] We now use notation (5.51) rather than $x_0 = \text{Rl}(Ae^{it})$ in order to avoid confusion in evaluating the nonlinear terms when it is important to note that $\text{Rl}(A^2) \neq (\text{Rl}A)^2$.

Hence, there will be "secular" terms of the form $te^{\pm it}$ in x_2 and these will grow with time so that our expansion for x will become invalid when $\varepsilon^2 t = O(1)$. Since the solution will inevitably contain oscillations on a timescale of $O(1)$, the procedure here is to introduce a *slow* time variable $T = \varepsilon^2 t$ and regard x as a function of *both* t and T. This is called the *method of multiple scales* and it is described in detail in Kevorkian and Cole [19] or Hinch [10]. In practice, all we need to do to use the method is to blithely use the chain rule to replace d/dt by $\partial/\partial t + \varepsilon^2(\partial/\partial T)$ so that equation (5.47) for x becomes

$$\frac{\partial^2 x}{\partial t^2} + 2\varepsilon^2 \frac{\partial^2 x}{\partial t \partial T} + \varepsilon^4 \frac{\partial^2 x}{\partial T^4} + x = \varepsilon x^2.$$

Now, expanding in powers of ε again, x_0 and x_1 take exactly the same forms (5.51) and (5.52) as before as long as we now regard A and B as functions of T. The equation for x_2 becomes

$$\frac{\partial^2 x_2}{\partial t^2} + x_2 = 2x_0 x_1 - \frac{2\partial^2 x_0}{\partial t \partial T}$$

and we can *eliminate* the offending terms in $e^{\pm it}$ on the right-hand side if we make A satisfy the equation

$$2i\frac{dA}{dT} = \frac{10}{3}A^2 A^*. \tag{5.53}$$

Solutions of this equation for A will give x_0 as a slowly modulated oscillatory function which is a valid asymptotic approximation for x for times of $O(\varepsilon^{-2})$ (Exercise 5.9).

Now, let us apply these ideas to the periodic Stokes wave trains that we considered in Chapter 3. We can obtain the non-dimensional form of the equations by putting $\delta = 1$ in (5.36)–(5.38), and for simplicity, we consider the case of infinitely deep water.

Schematically, we can write this system as

$$\frac{\partial^2 \phi}{\partial x^2} + \frac{\partial^2 \phi}{\partial z^2} = 0, \tag{5.54}$$

with

$$\frac{\partial^2 \phi}{\partial t^2} + \frac{\partial \phi}{\partial z} = \varepsilon Q \tag{5.55}$$

on $z = 0$, where Q is a power series in ε involving terms that are nonlinear in ϕ and η. The precise form of Q will be needed if the subsequent analysis is to be followed in detail, and a recipe for it is given in Exercise 5.11. However, readers who simply want the general gist do not need this information if they are prepared to trust the authors' calculations!

Taking the wavenumber k to be positive, the linear Stokes wave train for deep water can be written as

$$\phi_0 = (Ae^{i(kx-\omega t)} + A^* e^{-i(kx-\omega t)})e^{kz}, \tag{5.56}$$

where $\omega^2 = k$ and, exactly as in the above model problem, we see that when we put $\phi \sim \phi_0 + \varepsilon\phi_1 + \varepsilon^2\phi_2 + \cdots$, ϕ_2 will grow algebraically in time. Thus, the expansion is invalid when $\varepsilon^2 t = O(1)$. Because we are now dealing with functions of several variables, it turns out that we need to introduce the four "slow" variables $X_1 = \varepsilon x$, $X_2 = \varepsilon^2 x$, $T_1 = \varepsilon t$, and $T_2 = \varepsilon^2 t$, but, otherwise, we proceed as before. If we regard ϕ as a function of x, z, t, X_1, T_1, X_2, and T_2, the equation for ϕ_0 is unchanged except that A will now be a function of X_1, T_1, X_2, and T_2. However, the equation for ϕ_1 becomes

$$\frac{\partial^2\phi_1}{\partial x^2} + \frac{\partial^2\phi_1}{\partial z^2} = -2\frac{\partial^2\phi_0}{\partial x\partial X_1}, \tag{5.57}$$

with

$$\frac{\partial^2\phi_1}{\partial t^2} + \frac{\partial\phi_1}{\partial z} = -2\frac{\partial^2\phi_0}{\partial t\partial T_1} - 2\frac{\partial\phi_0}{\partial x}\frac{\partial^2\phi_0}{\partial x\partial t} - 2\frac{\partial\phi_0}{\partial z}\frac{\partial^2\phi_0}{\partial z\partial t} \tag{5.58}$$

on $z = 0$, and so

$$\phi_1 = \left(-iz\frac{\partial A}{\partial X_1}e^{i(kx-\omega t)} + iz\frac{\partial A^*}{\partial X_1}e^{-i(kx-\omega t)}\right)e^{kz}$$
$$+(Be^{i(kx-\omega t)} + B^*e^{-i(kx-\omega t)})e^{kz},$$

where B depends on X_1, T_1, X_2, and T_2. Using the boundary condition (5.58), we find that the elimination of terms in $e^{\pm i(kx-\omega t)}$ requires[8]

$$\frac{\partial A}{\partial X_1} + 2\omega\frac{\partial A}{\partial T_1} = 0.$$

Thus, A is a function of $X_1 - VT_1$, X_2, and T_2, where $V = \frac{1}{2\omega}$, and remembering that $\omega^2 = k$, we can identify V as the group velocity $d\omega/dk$. This observation points to a very general result in the theory of modulated linear wave trains that says that we can describe the gradual effect of nonlinearity if and only if we work in a frame moving with the group velocity.

Finally, we set $\xi = X_1 - VT_1$ and carry on to the next term in the expansion for ϕ. We find that

$$\phi_2 = \left(-\frac{1}{2}\frac{\partial^2 A}{\partial\xi^2}z^2 - i\left(\frac{\partial A}{\partial X_2} + \frac{\partial B}{\partial\xi}\right)z + C\right)e^{i(kx-\omega t)+kz} + \text{complex conjugate},$$

where C is a function of ξ, X_2, and T_2. When we apply the boundary condition (5.55) to $O(\varepsilon^2)$, we find that the terms in $e^{\pm i(kx-\omega t)}$ will balance only if A satisfies the *Nonlinear Schrödinger equation* (NLS equation)

$$i\left(\frac{\partial A}{\partial T_2} + V\frac{\partial A}{\partial X_2}\right) - V^3\frac{\partial^2 A}{\partial\xi^2} + cA^2A^* = 0, \tag{5.59}$$

---

[8] Note that if we had introduced the variable $T_1 = \varepsilon t$ in the solution of our model equation (5.47), we would merely have found that $\partial A/\partial T_1 = 0$.

where c is a real constant (see Exercise 5.11). This equation can be simplified by transforming to a frame moving with the group velocity on the T_2 scale and writing $\zeta = X_2 - VT_2$ and $\tau = T_2$, so that as a function of ξ, ζ, and τ, A satisfies

$$i\frac{\partial A}{\partial \tau} - V^3\frac{\partial^2 A}{\partial \xi^2} + cA^2A^* = 0, \tag{5.60}$$

which is the more familiar form of the NLS equation.

Despite its apparent complexity, this equation has been much studied and as much is known about it as is known about the KdV equation. Notice the following:

(i) If the dispersive term $\partial^2 A/\partial \xi^2$ is absent, the NLS equation effectively reduces to (5.53).
(ii) If the nonlinear terms are neglected, the NLS equation reduces to what is effectively the beam equation (Exercise 4.10).
(iii) There is a particular solution of the form

$$A = \left(-\frac{2V^3}{c}\right)^{1/2} e^{i\beta\zeta - iV^3\tau}\operatorname{sech}\xi, \tag{5.61}$$

where β is any real constant. This solution can be written in terms of (x, t) as

$$A = \left(-2\frac{V^3}{c}\right)^{1/2} e^{i\varepsilon^2(\beta x - (\beta V + V^3)t)}\operatorname{sech}(\varepsilon(x - Vt)),$$

which illustrates the modulation in phase and amplitude that governs the eventual fate of the Stokes wave train (5.56).

Although (5.61) can be likened to the solitary wave solution (5.46) of the KdV equation, we must remember that A is now the modulation of the wave train (5.56), so we have a so-called "envelope" solitary wave. However, there is a much more profound mathematical unity between (5.44) and (5.60) because both equations are suceptible to the theory of *inverse scattering* (Drazin and Johnson [20]. In fact, we have uncovered just the tip of an intellectual iceberg, and (5.46) and (5.61) have far more interesting attributes than simply being spatially localized solitary waves. Hence, they have earned the sobriquet of solitons, about which far more can be found in specialized texts such as Dodd et al. [21], which study *partial* differential equations that are *completely integrable* in the Hamiltonian sense. Both the KdV equation and the NLS equation are completely integrable and their solution can, in principle, be written down when appropriate data are prescribed at $\tau = 0$ for all ξ. A remarkable feature is that most reasonable initial data will lead to a universal behavior as $\tau \to \infty$ in which the solution will consist of a series of solitary waves of the form (5.46) or (5.61).[9] Their more specialized name solitons reflects the fact

---

[9] Note that we are, at the moment, unable to make any comparable statement about the relatively trivial (5.1).

that, for large times, they move like independent particles, even to the extent of "passing through" each other while retaining their own amplitude and velocity after each "collision." The theory of inverse scattering can predict the size and number of these solitons in any particular situation.

5.3 Smooth Solutions for Nonlinear Waves

We now present prototypical solutions of the three models described in Section 5.2, each of which relies on the existence of a region of *simple wave flow*. This concept describes any situation in which the region of interest is adjacent to a *uniform region* in which all of the flow variables are constant, and this enables us to exploit our knowledge of the Riemann invariants. The region of interest will be the simple wave region that is "spanned" by one family of characteristics that emanate from the uniform region. Thus, the corresponding Riemann invariant will take a known constant value *everywhere* in the region of interest. Hence, a problem with two scalar dependent variables is immediately reduced to a single scalar first-order partial differential equation which can be solved by traditional methods. To make things even easier, we can note that the constancy of the second Riemann invariant along characteristics of the second family implies that these characteristics will be straight lines and that the flow variables will be constant along these lines. We now use this strategy to obtain exact analytic solutions to three famous problems.

5.3.1 The Piston Problem for One-dimensional Unsteady Gasdynamics

We suppose that gas is at rest with speed of sound c_0 in $x < 0$ in a tube $-\infty < x < \infty$ containing a piston at $x = 0$. At $t = 0$, the piston begins to move in the positive x direction with given velocity $\dot{X}(t)$. The flow will be homentropic and (5.7) and (5.8) will therefore hold in $x < X(t)$ where $x = X(t)$, is the position of the piston at time t. We first assume that $\dot{X}(t)$ is a monotonically increasing function of t and $X(0) = \dot{X}(0) = 0$.

   In the region $x < 0$, $t < 0$, $u = 0$, and $c = c_0$, so that the characteristics there will be straight lines with slope $\pm c_0$. For $t > 0$, we first consider a point $P(x,t)$ as shown in Figure 5.2, which is such that it is at the intersection of two characteristics PA and PB emanating from the region $x < 0$, $t < 0$. Along PA, $u + 2c/(\gamma - 1) = 2c_0/(\gamma - 1)$, and along PB, $u - 2c/(\gamma - 1) = -2c_0/(\gamma - 1)$, so we can deduce straightaway that $u = 0$ and $c = c_0$ at P and that the characteristics PA and PB are straight lines with slope $\pm c_0$. Hence, P lies in a uniform region in which $u = 0$ and $c = c_0$ as long as $x < -c_0 t$; this is just another way of saying that the disturbance caused by the piston propagates into the gas with speed c_0.

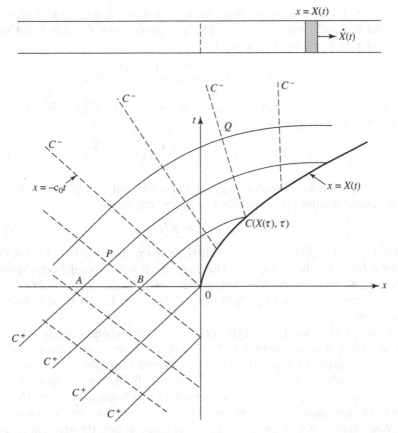

Fig. 5.2. The piston problem: — positive characteristics, − − − negative characteristics.

When $x > -c_0 t$, we can use the fact that all of the *positive* characteristics, labeled C^+ in Figure 5.2, emanate from the uniform region and so

$$u + \frac{2c}{\gamma - 1} = \frac{2c_0}{\gamma - 1} \qquad (5.62)$$

holds everywhere in $x < X(t)$. In particular, since we know that $u = \dot{X}(t)$ on the piston, we can deduce that $c = c_0 - (\gamma - 1)/2\dot{X}(t)$ on the piston. This relation only makes physical sense if $c \geq 0$, and so we will, for the moment, impose the further restriction that

$$\dot{X}(t) \leq \frac{2c_0}{\gamma - 1}.$$

As described earlier, the region $-c_0 t < x < X(t)$ is a region of simple wave flow, and if we consider the solution at the point Q in this region, we may use the fact that the negative characteristic through Q will be a straight line along which u and c are both constant. We suppose that C is the point where $t = \tau$ and $x = X(\tau)$, so that along QC,

$$u = \dot{X}(\tau) \quad \text{and} \quad c = c_0 - \frac{(\gamma - 1)}{2}\dot{X}(\tau). \tag{5.63}$$

Since QC is a negative characteristic, labeled C^- in Figure 5.2, on which $dx/dt = u - c$, the equation for QC is

$$x - X(\tau) = (t - \tau)\left(\frac{\gamma + 1}{2}\dot{X}(\tau) - c_0\right); \tag{5.64}$$

thus, the solution for u and c is given parametrically by (5.63) and (5.64). This solution is valid for all points x and t which satisfy

$$-c_0 t < x < X(t),$$

provided that $0 < \dot{X}(\tau) < 2c_0/(\gamma - 1)$. If $\dot{X}(\tau)$ exceeds $2c_0/(\gamma - 1)$, a vacuum will form between the piston and the gas, which will expand freely in such a way that it is bounded by the characteristic with slope $2c_0/(\gamma - 1)$ on which $c = 0$. The solution given by (5.63) and (5.64) will still be valid behind this characteristic.

By considering functions $X(t)$ such that $\ddot{X}(0)$ becomes larger and larger while $\dot{X}(t)$ tends to a constant V for $t > 0$, an analytic solution can also be found to the problem when the piston is started impulsively and moves out of the tube with speed $V \le 2c_0/(\gamma - 1)$. In this limit, the C^- characteristics more and more nearly fan out from the origin in Figure 5.2 and, in the limit, there is an *expansion fan* or *centered simple wave* between two regions of uniform flow, as shown in Figure 5.3. In the expansion fan, the straight negative characteristics all go through the origin and the solution in the fan is such that u and c depend only on x/t. This fact can also be deduced directly from a dimensional argument since there is no length scale in this problem, whose solution can therefore only involve the parameters V and c_0 (Exercise 5.15).

If the piston moves *into* the fluid, we immediately encounter a completely different scenario. The slope of the negative characteristics QC is such that these characteristics will inevitably intersect. This causes the above solution to break down because u will become multivalued, as described in Section 5.1. We can then only obtain a single-valued solution by introducing a discontinuity into u and c which is known as a shock wave and this will be discussed in Chapter 6.

We conclude with the following observation. Equations (5.9) are of the form $\partial r_i/\partial t + \lambda_i(\partial r_i/\partial x) = 0$, where r_1, and r_2 are the Riemann invariants and λ_1 and λ_2 are the slopes of the characteristics. These equations can be transformed by regarding r_1 and r_2 as independent variables and this results in the following *linear* set of equations for x and t:

$$\frac{\partial x}{\partial r_2} = \lambda_1 \frac{\partial t}{\partial r_2} \quad \text{and} \quad \frac{\partial x}{\partial r_1} = \lambda_2 \frac{\partial t}{\partial r_1}.$$

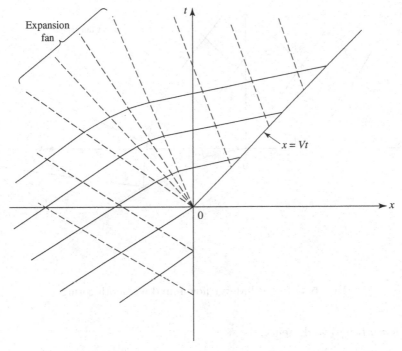

Fig. 5.3. Characteristics for the impulsively-withdrawn piston problem.

This formulation is of interest in connection with the hodograph transformation described in Section 5.4 and can also be used to solve the problem of intersecting simple waves (Exercise 5.19).

5.3.2 Prandtl–Meyer Flow

The next, slightly less simple, example concerns the flow of a two-dimensional steady supersonic flow past a continuous convex corner. We suppose that a uniform supersonic flow with Mach number M_1 flows parallel to a wall along $y = 0$ in $x < 0$ and that the corner starts smoothly at $x = 0$, as shown in Figure 5.4. The characteristic picture that emerges is exactly analogous to the piston problem of Section 5.3.1 where now we have to use (5.20) and (5.21) in place of (5.10).

   In the incoming flow, $\mu = \mu_1 = \sin^{-1}(1/M_1)$ and $\theta = 0$. Thus, the characteristics are straight lines making angles $\pm\mu_1$ with the x axis, and by the arguments used in Section 5.3.1, the influence of the corner will not be felt in the uniform region upstream of the characteristic $y = x \tan \mu_1$ through the origin. Again, we can see that the negative characteristics all emanate from the undisturbed region into the simple wave region between the curved wall

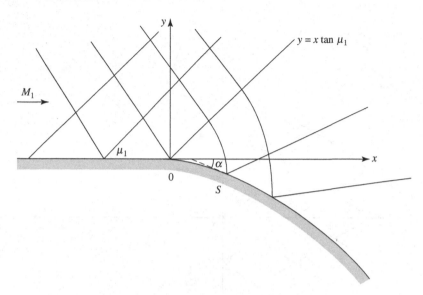

Fig. 5.4. Prandtl-Meyer flow round a smooth corner.

and $y = x \tan \mu_1$ and, hence,

$$\theta + f(\mu) = f(\mu_1) \tag{5.65}$$

there, where, from (5.20),

$$f(\mu) = \mu + \frac{1}{\lambda} \tan^{-1}(\lambda \cot \mu), \tag{5.66}$$

and, as usual, $\lambda^2 = (\gamma - 1)/(\gamma + 1)$. At a point S on the wall where the slope of the wall is $-\alpha$, we know that $\theta = -\alpha$ and so μ is determined from (5.65). Moreover, from (5.66), we can see that $f(\mu)$ is a decreasing function of μ with a maximum of $\pi/2\lambda$ at $\mu = 0$ and a minimum of $\pi/2$ at $\mu = \pi/2$, as shown in Figure 5.5. Hence, the maximum angle through which the flow can be turned will be $\pi/2\lambda - \pi/2$ and this can only occur when μ_1 is $\pi/2$ and the incoming flow is sonic. It is also clear that μ decreases as the flow turns around the corner, and since M increases as μ decreases, this means that as long as our assumption that the incoming flow is supersonic is valid, the flow will stay supersonic throughout. As in Section 5.3.1, we may use the fact that the positive characteristic through S is a straight line along which θ and μ are constant to write down enough equations to determine the flow everywhere (Exercise 5.16).

There is one case in which we can make easy analytical progress and that is for the supersonic flow past a sharp corner which turns the flow through an angle α, as illustrated in Figure 5.6. Now, by analogy with the argument at the end of Section 5.3.1, the flow consists of a *Prandtl–Meyer expansion*

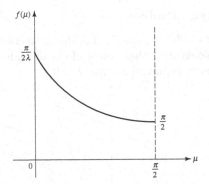

Fig. 5.5. The function $f(\mu)$.

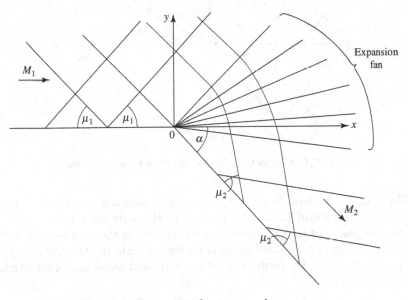

Fig. 5.6. Supersonic flow past a sharp corner.

fan in which the positive characteristics are straight lines through the origin separating a uniform region where $\theta = 0$ and $\mu = \mu_1$ from a second uniform region where $\theta = -\alpha$ and $\mu = \mu_2$, where $f(\mu_2) = \alpha + f(\mu_1)$. The details of the flow in the expansion fan are left to Exercise 5.17.

We note that if the corner is concave, we will once again encounter the problem of intersecting characteristics and we will address this difficulty in Chapter 6.

5.3.3 The Dam Break Problem

For the shallow water model, a paradigm problem to consider is that of a dam breaking suddenly. We assume that water of depth h_0 is retained in $x < 0$ by a dam which is suddenly removed at time $t = 0$.

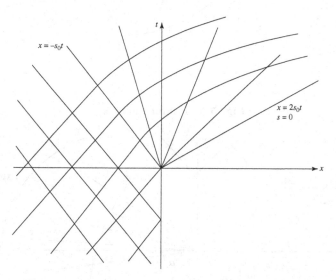

Fig. 5.7. Characteristics for the dam break problem.

Using the fact that, from (5.31), $u \pm 2s$ is constant on $dx/dt = u \pm s$, we see that there will be no disturbance in the uniform region $x < -s_0 t$, where $s_0^2 = gh_0$, and that $u + 2s = 2s_0$ everywhere in the simple wave region, as shown in Figure 5.7. Since there is no length scale in this problem, there will be an expansion fan centered on the origin, and along each characteristic through the origin,

$$\frac{x}{t} = u - s.$$

Hence,

$$u = \frac{2}{3}\left(\frac{x}{t} + s_0\right) \quad \text{and} \quad s = \frac{1}{3}\left(2s_0 - \frac{x}{t}\right) \tag{5.67}$$

within the fan, and since the depth of the water vanishes when $s = 0$, this solution will only hold for

$$-s_0 t < x < 2s_0 t.$$

The depth of the water at time t is shown in Figure 5.8.

This solution is analogous to the instantaneous removal of the piston in Section 5.3.1. An obvious extension would be to remove a dam between two reservoirs containing water of different heights, but a quick sketch of the characteristics shows that, inevitably, some positive characteristics will intersect each other and so this problem will also be deferred to Chapter 6.

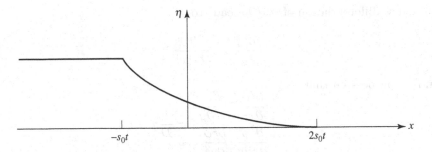

Fig. 5.8. The water depth in the dam break problem.

5.4 *The Hodograph Transformation

In Section 5.2.2, we have already commented on the difficulty of solving the equations of two-dimensional steady flow, even in the irrotational case. However, since the coefficients in (5.17) are only functions of the velocity components u and v, the idea of working in the *hodograph plane*, in which u and v are the independent variables, suggests itself as a possible route to a linear model. If we were simply to regard ϕ as a function of these variables, we would make little progress and instead we first make the *Legendre transformation* to the new variable

$$\psi(u, v) = xu + yv - \phi(x, y), \tag{5.68}$$

where x and y are now regarded as functions of u and v. There are good geometric reasons (see Ockendon et al. [9]) for making this transformation and we will see that it results in a linear equation for ψ. From (5.68), we find that

$$\frac{\partial \psi}{\partial u} = x + \left(u \frac{\partial x}{\partial u} + v \frac{\partial y}{\partial u} - \frac{\partial \phi}{\partial u} \right) = x$$

and, similarly, $\partial \psi / \partial v = y$. Then,

$$1 = \frac{\partial}{\partial u} \left(\frac{\partial \phi}{\partial x} \right) = \frac{\partial^2 \phi}{\partial x^2} \frac{\partial^2 \psi}{\partial u^2} + \frac{\partial^2 \phi}{\partial x \partial y} \frac{\partial^2 \psi}{\partial u \partial v}$$

and

$$0 = \frac{\partial}{\partial v} \left(\frac{\partial \phi}{\partial x} \right) = \frac{\partial^2 \phi}{\partial x^2} \frac{\partial^2 \psi}{\partial u \partial v} + \frac{\partial^2 \phi}{\partial x \partial y} \frac{\partial^2 \psi}{\partial v^2}.$$

Hence,

$$\frac{\partial^2 \phi}{\partial x^2} = \frac{1}{D} \frac{\partial^2 \psi}{\partial v^2} \quad \text{and} \quad \frac{\partial^2 \phi}{\partial x \partial y} = -\frac{1}{D} \frac{\partial^2 \psi}{\partial u \partial v},$$

where

$$D = \begin{vmatrix} \dfrac{\partial^2 \psi}{\partial u^2} & \dfrac{\partial^2 \psi}{\partial u \partial v} \\[2ex] \dfrac{\partial^2 \psi}{\partial u \partial v} & \dfrac{\partial^2 \psi}{\partial v^2} \end{vmatrix}.$$

Similarly, differentiation of $\partial\phi/\partial y$ leads to

$$\frac{\partial^2\phi}{\partial y^2} = \frac{1}{D}\frac{\partial^2\psi}{\partial u^2}$$

and it can be seen that

$$\begin{vmatrix} \dfrac{\partial^2\phi}{\partial x^2} & \dfrac{\partial^2\phi}{\partial x\partial y} \\[2mm] \dfrac{\partial^2\phi}{\partial x\partial y} & \dfrac{\partial^2\phi}{\partial y^2} \end{vmatrix} = \frac{1}{D}.$$

This transformation is valid as long as D is bounded and non-zero.

Eventually, (5.17) becomes

$$(c^2 - u^2)\frac{\partial^2\psi}{\partial v^2} + 2uv\frac{\partial^2\psi}{\partial u\partial v} + (c^2 - v^2)\frac{\partial^2\psi}{\partial u^2} = 0, \qquad (5.69)$$

where $c^2 = c_0^2 - [(\gamma - 1)/2](u^2 + v^2)$. This equation is linear and it can be transformed into an even simpler form by writing $u = q\cos\theta$ and $v = q\sin\theta$ to get

$$\frac{\partial^2\psi}{\partial\theta^2} + \frac{q^2 c^2}{c^2 - q^2}\frac{\partial^2\psi}{\partial q^2} + q\frac{\partial\psi}{\partial q} = 0, \qquad (5.70)$$

where

$$c^2 = c_0^2 - \frac{\gamma - 1}{2}q^2;$$

this is known as *Chaplygin's equation* and it is clearly susceptible to separation of variables. Chaplygin's equation can be shown to be equivalent to the equations obtained by using the Riemann invariants as independent variables, as described at the end of Section 5.3.1.

Unfortunately, two factors limit the usefulness of (5.70). The first is the condition that D should be bounded and non-zero. In uniform flow, for example, $\psi = 0$ and the flow region maps into a single point on the hodograph plane because $D = 0$. Similarly, for simple wave flow, one of the Riemann invariants is constant and the flow region maps into a single characteristic curve in the hodograph plane.

Second, and more crippling, is the fact that everyday boundary conditions in the physical plane can become unmanageable in the hodograph plane. Suppose, for example, there is a fixed boundary $y = f(x)$; then, the boundary condition is

$$\frac{\partial\psi}{\partial v} = f\left(\frac{\partial\psi}{\partial u}\right) \quad \text{on } v = uf'\left(\frac{\partial\psi}{\partial u}\right),$$

and unless the boundary is straight, the nonlinearity reappears in the problem via the boundary conditions.

Exercises

R5.1 Suppose that $\partial u/\partial t + u\partial u/\partial x = 0$. Show that if $u(x,0) = u_0(x)$, then $u = u_0(x - ut)$.

When $u_0 = 1/(1 + x^2)$, sketch the evolution of u as t increases by translating the graph of u_0 in the x direction by a distance that depends on u and assuming the graph remains smooth. Show that the characteristics are $x = t/(1 + s^2) + s$, where s is a parameter, and that on the envelope of these characteristics, $2ts = (1 + s^2)^2$. Deduce that $\partial u/\partial x$ first becomes infinite at $x = \sqrt{3}$ and $t = 8\sqrt{3}/9$.

5.2 From (5.14) and (5.16), show that for any function λ,

$$\left[(c^2 - u^2)\frac{\partial}{\partial x} + (\lambda - uv)\frac{\partial}{\partial y}\right]u + \left[(-uv - \lambda)\frac{\partial}{\partial x} + (c^2 - v^2)\frac{\partial}{\partial y}\right]v = 0.$$

Show that the two differential operators in this equation are proportional to each other when $\lambda = \pm c\sqrt{u^2 + v^2 - c^2}$ and deduce (5.18) and (5.19).

Now write $u = cr\cos\theta$ and $v = cr\sin\theta$ and show that the slopes of the characteristics are

$$\frac{dy}{dx} = \frac{-r^2\sin\theta\cos\theta \pm \sqrt{r^2 - 1}}{1 - r^2\cos^2\theta},$$

and then set $r = \operatorname{cosec}\mu$ to derive the equation of the characteristics in the form (5.21).

[If you feel strong, you can now go on to check that the Riemann invariants are given by (5.20).]

5.3 Show that if the base of a shallow stream is $z = -b(x)$, then the two-dimensional shallow water equations are

$$\frac{\partial\eta}{\partial t} + \frac{\partial}{\partial x}(u(\eta + b)) = 0,$$

$$\frac{\partial u}{\partial t} + u\frac{\partial u}{\partial x} = -g\frac{\partial\eta}{\partial x}.$$

Deduce that if $b = mx$, then

$$u \pm 2s - mgt = \text{constant on} \quad \frac{dx}{dt} = u \pm s,$$

where $s^2 = g(\eta + mx)$.

Suppose the initial conditions are $u = F(x)$ and $s = s_0 - \frac{1}{2}F(x)$ at $t = 0$. Show that the flow is a simple wave flow with $u + 2s - mgt = 2s_0$ and that

$$u - mgt = F\left(x - \frac{3}{2}ut + s_0 t + mgt^2\right).$$

5.4 The scaling used to obtain (5.22)–(5.24) shows that for two-dimensional waves on shallow water, the equations reduce to

$$\frac{\partial u}{\partial x} + \frac{\partial w}{\partial z} = 0,$$

$$\frac{\partial u}{\partial t} + u\frac{\partial u}{\partial x} + w\frac{\partial u}{\partial z} = -\frac{1}{\rho}\frac{\partial p}{\partial x},$$

$$\frac{\partial p}{\partial z} = -g\rho.$$

Show that the *rotational* flow $u = u_0(z)$, $w = 0$, $\eta = \eta_0 = $ constant satisfies these equations. Show also that in tidal waves for which $\bar{u} = u - u_0$, w, and $\bar{\eta} = \eta - \eta_0$ are small,

$$\frac{\partial \bar{u}}{\partial x} + \frac{\partial w}{\partial z} = 0,$$

$$\frac{\partial \bar{u}}{\partial t} + u_0\frac{\partial \bar{u}}{\partial x} + u_0'w = -g\frac{\partial \bar{\eta}}{\partial x},$$

with $w = 0$ on $z = 0$ and

$$\frac{\partial \bar{\eta}}{\partial t} + u_0\frac{\partial \bar{\eta}}{\partial x} = w \quad \text{on } z = \eta_0.$$

Now, suppose that $\bar{\eta} = \mathrm{Rl}(ae^{ik(x-ct)})$, where a, k, and c are constants. Show that $w = \mathrm{Rl}(f(z)e^{ik(x-ct)})$, where

$$(u_0 - c)f' - u_0'f = igak,$$

with $f(0) = 0$, and $f(\eta_0) = iak(u_0(\eta_0) - c)$. Hence, show that c satisfies

$$g\int_0^{\eta_0} \frac{dz}{(c - u_0(z))^2} = 1.$$

Note that this calculation shows that if there are values of z for which $c = u_0(z)$, then nonlinear terms need to be retained locally near these values of z, which are called *critical layers*.

5.5 Show that for two-dimensional flow, the shallow water equations (5.27) and (5.30) generalize to

$$\frac{\partial u}{\partial t} + u\frac{\partial u}{\partial x} + v\frac{\partial u}{\partial y} = -g\frac{\partial \eta}{\partial x},$$

$$\frac{\partial v}{\partial t} + u\frac{\partial v}{\partial x} + v\frac{\partial v}{\partial y} = -g\frac{\partial \eta}{\partial y},$$

$$\frac{\partial \eta}{\partial t} + \frac{\partial}{\partial x}(u\eta) + \frac{\partial}{\partial y}(v\eta) = 0.$$

Show further that in steady flow, $\frac{1}{2}(u^2 + v^2) + g\eta$ is conserved along a streamline and that with $\mathbf{u} = (u, v, 0)$,

$$\mathbf{u} \wedge (\nabla \wedge \mathbf{u}) = \nabla\left(\tfrac{1}{2}(u^2 + v^2) + g\eta\right).$$

Compare this result with Crocco's theorem (Exercise 2.5) and show that if $\frac{1}{2}(u^2 + v^2) + g\eta$ varies from streamline to streamline, then $\nabla \wedge \mathbf{u} \neq \mathbf{0}$.

*5.6 Three-dimensional waves on shallow water are modeled by

$$\frac{\partial^2 \phi}{\partial z^2} + \delta^2 \left(\frac{\partial^2 \phi}{\partial x^2} + \frac{\partial^2 \phi}{\partial y^2} \right) = 0,$$

with

$$\frac{\partial \phi}{\partial t} + \eta + \frac{\varepsilon}{2\delta^2}\left(\left(\frac{\partial \phi}{\partial z}\right)^2 + \delta^2\left(\left(\frac{\partial \phi}{\partial x}\right)^2 + \left(\frac{\partial \phi}{\partial y}\right)^2 \right) \right) = 0$$

and

$$\frac{\partial \phi}{\partial z} = \delta^2 \frac{\partial \eta}{\partial t} + \varepsilon\delta^2 \left(\frac{\partial \phi}{\partial x}\frac{\partial \eta}{\partial x} + \frac{\partial \phi}{\partial y}\frac{\partial \eta}{\partial y} \right)$$

on $z = \varepsilon\eta(x, 0)$, and

$$\frac{\partial \phi}{\partial z} = 0 \quad \text{on } z = -1.$$

(i) Suppose that δ and ε are both small and that $\delta^2 \ll O(\varepsilon)$. Suppose also that the waves only vary gradually in the y direction, with a length scale $\varepsilon^{-1/2}$. Show that over times of $O(\varepsilon^{-1})$, the surface elevation satisfies the "dispersionless Kadomtsev–Petviashvili equation"

$$\frac{\partial}{\partial \xi}\left(\frac{\partial \eta}{\partial \tau} + \frac{3}{2}\eta\frac{\partial \eta}{\partial \xi} \right) + \frac{1}{2}\frac{\partial^2 \eta}{\partial Y^2} = 0,$$

where $Y = \sqrt{\varepsilon}y$, $\tau = \varepsilon t$, and $\xi = x - t$.

(ii) When $\delta^2 = \kappa\varepsilon$ and κ is of $O(1)$, deduce that (5.42) and (5.43) generalize to

$$\frac{\partial \phi_0}{\partial t} + \eta_0 + \frac{1}{2}\varepsilon\left(\left(\frac{\partial \phi_0}{\partial x}\right)^2 + \left(\frac{\partial \phi_0}{\partial y}\right)^2 \right) + \kappa\varepsilon\left(\frac{\partial A}{\partial t} + \eta_1 \right) = O(\varepsilon^2)$$

and

$$\frac{\partial \eta_0}{\partial t} + \frac{\partial^2 \phi_0}{\partial x^2} + \frac{\partial^2 \phi_0}{\partial y^2} + \varepsilon\left[\eta_0\left(\frac{\partial^2 \phi_0}{\partial x^2} + \frac{\partial^2 \phi_0}{\partial y^2} \right) + \frac{\partial \phi_0}{\partial x}\frac{\partial \eta_0}{\partial x} + \frac{\partial \phi_0}{\partial y}\frac{\partial \eta_0}{\partial y} \right.$$
$$\left. + \kappa\left(\frac{\partial^2}{\partial x^2} + \frac{\partial^2}{\partial y^2} \right)\left(A + \frac{1}{3}\left(\frac{\partial^2 \phi_0}{\partial x^2} + \frac{\partial^2 \phi}{\partial y^2} \right) \right) + \kappa\frac{\partial \eta_1}{\partial t} \right] = O(\varepsilon^2),$$

respectively.

Now assume the wave motion is in the x direction to lowest order, so that ϕ_0 and η_0 are functions of $x - t = \xi$, $\tau = \varepsilon t$, and $Y = \sqrt{\varepsilon} y$. Show that, as in the two-dimensional case,

$$\frac{\partial \phi_0}{\partial \xi} = \eta_0,$$

and, hence, that the terms of $O(\varepsilon)$ give that ϕ_0 and η_0 also satisfy

$$\frac{\partial \eta_0}{\partial \tau} + \frac{3}{2} \eta_0 \frac{\partial \eta_0}{\partial \xi} + \frac{1}{6} \frac{\partial^3 \eta_0}{\partial \xi^3} + \frac{1}{2} \frac{\partial^2 \phi_0}{\partial Y^2} = 0.$$

This is called the "Kadomtsev–Petviashvili model" (Drazin and Johnson [20]).

5.7 Set $\varepsilon = 1$ in (5.37)–(5.39) and show that as $\delta \to 0$,

$$\frac{\partial \phi_0}{\partial t} + \eta_0 + \frac{1}{2} \left(\frac{\partial \phi_0}{\partial x} \right)^2 = 0$$

and

$$\frac{\partial^2 \phi_0}{\partial x^2} + \frac{\partial \eta_0}{\partial t} + \frac{\partial \phi_0}{\partial x} \frac{\partial \eta_0}{\partial x} + \eta_0 \frac{\partial^2 \phi_0}{\partial x^2} = 0,$$

using the notation of (5.40). Show that these equations are equivalent to the shallow water equations (5.27) and (5.30).

Note that in the derivation of (5.27) and (5.30), no explicit assumption was made about irrotationality and yet the above method relies on (5.36), which assumes irrotationality. From (5.28), we see that $\omega = \nabla \wedge \mathbf{u} = (0, 0, -z \frac{\partial^2 u}{\partial x^2})$, and using the scalings of (5.36), $|\omega| = O(\delta^2 \sqrt{\frac{g}{h}})$. Hence, in the shallow water approximation, the vorticity may be taken to be zero.

*5.8 Immiscible fluids of density ρ_w and ρ_0 (perhaps water and oil) flow along a horizontal channel $0 < y < D$ with the oil above the water. The interface is at $y = h(x, t)$. Making the shallow water assumptions in each fluid, show that the horizontal velocities u_w and u_0 satisfy

$$\frac{\partial h}{\partial t} + \frac{\partial}{\partial x}(h u_w) = 0,$$

$$-\frac{\partial h}{\partial t} + \frac{\partial}{\partial x}((D - h) u_0) = 0,$$

$$\frac{\partial u_w}{\partial t} + u_w \frac{\partial u_w}{\partial x} + g \frac{\partial h}{\partial x} + \frac{1}{\rho_w} \frac{\partial p}{\partial x} = 0,$$

and

$$\frac{\partial u_0}{\partial t} + u_0 \frac{\partial u_0}{\partial x} + g \frac{\partial h}{\partial x} + \frac{1}{\rho_0} \frac{\partial p}{\partial x} = 0,$$

where $p(x,t)$ is the pressure at the interface. Show that for small disturbances about the uniform state $h = H$, $u_w = u_0 = U$, and $p = p_0$, the
perturbation $\bar h = h - H$ satisfies

$$\left(\frac{\rho_0}{D-H} + \frac{\rho_w}{H}\right)\left(\frac{\partial^2 \bar h}{\partial t^2} + 2U\frac{\partial^2 \bar h}{\partial x \partial t} + U^2\frac{\partial^2 \bar h}{\partial x^2}\right) + g(\rho_0 - \rho_w)\frac{\partial^2 \bar h}{\partial x^2} = 0.$$

Deduce that waves can propagate on the interface as long as $\rho_0 < \rho_w$.

5.9 With $A = re^{i\theta}$ in (5.53), show that

$$r = r_0 = \text{constant},$$
$$\theta = -\frac{5}{3}r_0^2 T + \text{constant}.$$

Deduce that the effect of the nonlinearity over timescales of $t = O(\varepsilon^{-2})$ is
to change the period of the solution (whose amplitude is given by $r = r_0$)
from 2π (as predicted by (5.51)) to $2\pi/(1 - \frac{5}{3}\varepsilon^2 r_0^2)$.

Confirm that the solution is, in fact, periodic over all time scales when
$r_0 = O(1)$ by sketching the phase plane of (5.47), for which the phase
curves are

$$y^2 + x^2 = \frac{2}{3}\varepsilon x^3 + \text{constant},$$

where $y = dx/dt$.

*5.10 Show that in axes $\xi = x - c_0 t$ moving with the speed of sound in a station-
ary gas, the equations for the one-dimensional flow of a heat conducting
gas are

$$\frac{\partial \rho}{\partial t} + (u - c_0)\frac{\partial \rho}{\partial \xi} + \rho\frac{\partial u}{\partial \xi} = 0,$$
$$\rho\frac{\partial u}{\partial t} + \rho(u - c_0)\frac{\partial u}{\partial \xi} + \frac{\partial p}{\partial \xi} = 0$$

and

$$\frac{\partial p}{\partial t} + (u - c_0)\frac{\partial p}{\partial \xi} + \gamma p\frac{\partial u}{\partial \xi} = \frac{kR}{c_v}\frac{\partial^2 T}{\partial \xi^2},$$

in the usual notation. To study the long-time evolution of a small-
amplitude wave, write $t = \varepsilon^{-1}\tau$, $u \sim \varepsilon u_1 + \varepsilon^2 u_2 + \cdots$, $p \sim p_0 + \varepsilon p_1 + \varepsilon^2 p_2 + \cdots$, $\rho \sim \rho_0 + \varepsilon \rho_1 + \varepsilon^2 \rho_2 + \cdots$, and $T \sim T_0 + \varepsilon T_1 + \cdots$, where $p_0 = \rho_0 R T_0$.
Then, assuming $kR/c_v = \varepsilon \bar k$, derive the system

$$-c_0\frac{\partial \rho_1}{\partial \xi} + \rho_0\frac{\partial u_1}{\partial \xi}$$
$$+\varepsilon\left[\frac{\partial \rho_1}{\partial \tau} + u_1\frac{\partial \rho_1}{\partial \xi} + \rho_1\frac{\partial u_1}{\partial \xi} - c_0\frac{\partial \rho_2}{\partial \xi} + \rho_0\frac{\partial u_2}{\partial \xi}\right] = O(\varepsilon^2),$$
$$-c_0\rho_0\frac{\partial u_1}{\partial \xi} + \frac{\partial p_1}{\partial \xi}$$

$$+\varepsilon \left[\rho_0 \frac{\partial u_1}{\partial \tau} + \rho_0 u_1 \frac{\partial u_1}{\partial \xi} - c_0 \rho_1 \frac{\partial u_1}{\partial \xi} - c_0 \rho_0 \frac{\partial u_2}{\partial \xi} + \frac{\partial p_2}{\partial \xi}\right] = O(\varepsilon^2),$$

$$-c_0 \frac{\partial p_1}{\partial \xi} + \gamma p_0 \frac{\partial u_1}{\partial \xi}$$

$$+\varepsilon \left[\frac{\partial p_1}{\partial \tau} + u_1 \frac{\partial p_1}{\partial \xi} + \gamma p_1 \frac{\partial u_1}{\partial \xi} - c_0 \frac{\partial p_2}{\partial \xi} + \gamma p_0 \frac{\partial u_2}{\partial \xi} - \bar{k} \frac{\partial^2 T_1}{\partial \xi^2}\right] = O(\varepsilon^2).$$

Deduce that $\rho_1 = (\rho_0/c_0)u_1$, $p_1 = \rho_0 c_0 u_1$, and $p_1 = R(\rho_0 T_1 + T_0 \rho_1)$, and hence combine these equations and eliminate p_1, ρ_1 to obtain

$$2\rho_0 c_0 \frac{\partial u_1}{\partial \tau} + (\gamma+1)\rho_0 c_0 u_1 \frac{\partial u_1}{\partial \xi} = \bar{k} \frac{\partial^2 T_1}{\partial \xi^2} = \frac{\bar{k} T_0}{c_0}(\gamma-1) \frac{\partial^2 u_1}{\partial \xi^2}.$$

This equation can easily be reduced to *Burgers' equation* (see later, viz. (6.11) of Chapter 6). Unfortunately, the assumption that thermal conduction can be retained while viscosity is ignored is not true for a gas like air, but the effect of viscosity is merely to change the coefficients in Burgers' equation.

*5.11 (i) Take $\delta = 1$ and show that (5.37) and (5.38) can then be written as

$$\frac{\partial \phi}{\partial t} + \eta + \varepsilon \left[\eta \frac{\partial^2 \phi}{\partial z \partial t} + \frac{1}{2}\left(\frac{\partial \phi}{\partial x}\right)^2 + \frac{1}{2}\left(\frac{\partial \phi}{\partial z}\right)^2\right] = O(\varepsilon^2)$$

and

$$\frac{\partial \phi}{\partial z} - \frac{\partial \eta}{\partial t} + \varepsilon \left[\eta \frac{\partial^2 \phi}{\partial z^2} - \frac{\partial \phi}{\partial x} \frac{\partial \eta}{\partial x}\right] = O(\varepsilon^2)$$

on $z = 0$. Deduce that, to lowest order, Q in (5.55) is given by

$$-2 \left(\frac{\partial \phi}{\partial x} \cdot \frac{\partial^2 \phi}{\partial x \partial t} + \frac{\partial \phi}{\partial z} \cdot \frac{\partial^2 \phi}{\partial z \partial t}\right)$$

in accordance with (5.58).

(ii) Show that when the terms on the left-hand side of (5.55) are expanded in terms of the variables given before (5.57), the term of $O(\varepsilon^2)$ will be

$$\varepsilon^2 \left(\frac{\partial^2 \phi_2}{\partial t^2} + \frac{\partial \phi_2}{\partial z} + 2\frac{\partial^2 \phi_1}{\partial t \partial T_1} + \frac{\partial^2 \phi_0}{\partial T_1^2} + 2\frac{\partial^2 \phi_0}{\partial t \partial T_2}\right),$$

all evaluated on $z = 0$. Use the formulas for ϕ_0, and ϕ_1, and ϕ_2 that are given between (5.56) and (5.59) to show that the terms in this expression that are proportional to $e^{i(kx-\omega t)}$ are

$$\left(-i\frac{\partial A}{\partial X_2} - i\frac{\partial B}{\partial \xi} + i\frac{\partial B}{\partial \xi} + V^2 \frac{\partial^2 A}{\partial \xi^2} - 2i\omega \frac{\partial A}{\partial T_2}\right) e^{i(kx-\omega t)}.$$

Unfortunately, there are 19 nonlinear terms of $O(\varepsilon^2)$ which are proportional to $e^{i(kx-\omega t)}$, but following the pattern of (i), it is straightforward (but tedious) to see that they are all cubic in ϕ_0 and involve two t derivatives and three x or z derivatives. Hence, they are all proportional to $\omega^2 k^3 A^2 A^* e^{i(kx-\omega t)}$. Use this information to deduce (5.59).

*5.12 Show that if $\phi(x,t)$ satisfies (5.34) with $\phi(x,0) = \phi_0(x)$, $-\infty < x < \infty$, then its Fourier transform $\bar{\phi} = \int_{-\infty}^{\infty} \phi(x,t)e^{ikx}\, dx$ is

$$\bar{\phi}(k,t) = \bar{\phi}_0(k)e^{(ik-\varepsilon k^2)t}.$$

Deduce that as $\varepsilon \to 0$ for $x, t = O(1)$,

$$\phi = \frac{1}{2\pi}\int_{-\infty}^{\infty} \bar{\phi}_0(k)e^{ik(t-x)-\varepsilon k^2 t}\, dk \sim \phi_0(x-t).$$

However, show that when $t = \tau/\varepsilon$, where $\tau = O(1)$ and $\xi = x - t + O(1)$,

$$\phi = \frac{e^{-\xi^2/4\tau}}{2\pi}\int_{-\infty}^{\infty}\bar{\phi}_0(k)e^{-\tau(k+i\xi/2\tau)^2}\, dk.$$

When $\bar{\phi}_0(k)$ is well behaved at $k = -i\xi/2\tau$, show that

$$\phi \sim \frac{\text{constant}}{\sqrt{\tau}}e^{-\xi^2/4\tau}$$

as $\tau \to \infty$. To what initial condition for (5.35) does this solution correspond?

R5.13 The equations

$$\left(\frac{\partial}{\partial t} + (u \pm c)\frac{\partial}{\partial x}\right)\left(u \pm \frac{2c}{\gamma - 1}\right) = 0$$

for the gas velocity u and sound speed c are used to model gas flow in a tube under the action of a piston at $x = X(t)$. The gas is in $x < X(t)$ and $X(0) = 0$, $\dot{X}(t) \geq 0$ and $\ddot{X}(t) \geq 0$. When the gas is initially at rest with $c = c_0$, show that

$$u = \dot{X}(\tau),$$

where

$$u + \frac{2c}{\gamma - 1} = \frac{2c_0}{\gamma - 1} \quad \text{and} \quad \frac{x - X(\tau)}{t - \tau} = u - c$$

in the region $-c_0 t < x < X(t)$. Deduce the following:

(i) When $\dot{X}$ is a constant greater than $2c_0/(\gamma - 1)$, the gas expands into the region

$$-c_0 < \frac{x}{t} < \frac{2c_0}{\gamma - 1}.$$

(ii) When $X = gt^2/2$, then

$$\gamma u = \left(c_0 + \frac{\gamma + 1}{2}gt\right) - \left[\left(c_0 + \frac{\gamma + 1}{2}gt\right)^2 - 2\gamma g(c_0 t + x)\right]^{1/2}$$

in the region $-c_0 < x/t < gt/2$, for $t < 2c_0/(\gamma - 1)g$.

R5.14 Suppose that in the piston problem 5.13, the piston path is $x = -\frac{1}{2}gt^2$ where $g > 0$. Show that the negative characteristics are

$$\frac{x + \frac{1}{2}g\tau^2}{t - \tau} = -c_0 - \frac{\gamma + 1}{2}g\tau$$

and deduce that, for small τ, these characteristics form an envelope at

$$x = -c_0 t, \qquad t = 2c_0/(\gamma + 1)g.$$

Verify that $\partial u/\partial x$ is infinite at this point.

R5.15 A piston is withdrawn impulsively from a tube containing gas in $x < 0$. The model of Section 5.3 is

$$\left(\frac{\partial}{\partial t} + (u \pm c)\frac{\partial}{\partial x}\right)\left(u \pm \frac{2c}{\gamma - 1}\right) = 0$$

with $u = 0$, $c = c_0 = $ constant for $t = 0$, $x < 0$, and $u = V$ on $x = Vt$ for $t > 0$, with $V > 0$.

Taking L to be an *arbitrary* length scale, non-dimensionalize this model by writing

$$x = Lx', \qquad t = \frac{L}{Vt'}, \qquad u = Vu', \qquad c = c_0 c'$$

to give

$$\left(\frac{\partial}{\partial t'} + (Mu' \pm c')\frac{\partial}{\partial x'}\right)\left(Mu' \pm \frac{2c'}{\gamma - 1}\right) = 0,$$

where $M = V/c_0$, and

$$u' = 0, \quad c' = 1 \quad \text{for } t' = 0, \quad x' = 0,$$
$$u' = 1 \quad \text{on } x' = t' \text{ for } t' > 0.$$

Now, use the fact that u' is evidently only a function of x', t', and M to deduce that u/V is only a function of x/Vt and M.

Deduce that the solution is

$$u = \begin{cases} 0 & \text{if } \frac{x}{Vt} < -\frac{1}{M}, \\ \frac{2V}{\gamma+1}\left(\frac{x}{Vt} + \frac{1}{M}\right) & \text{if } -\frac{1}{M} < \frac{x}{Vt} < \frac{\gamma+1}{2} - \frac{1}{M}, \\ V & \text{if } \frac{\gamma+1}{2} - \frac{1}{M} < \frac{x}{Vt} < 1. \end{cases} \qquad (*)$$

Confirm this result by using (5.63) and (5.64) and noting that only small values of τ are relevant in generating the expansion fan. Use this idea to show that

$$x = t\left[\frac{\gamma + 1}{2}\dot{X}(\tau) - c_0\right]$$

and, hence, retrieve the solution $(*)$.

R5.16 With reference to Figure 5.4, show that on any negative characteristic,

$$\theta + f(\mu) = f(\mu_1),$$

where f is defined by (5.66). Show also that at the point S where the flow deflection is $-\alpha$, μ is given by

$$f(\mu) = f(\mu_1) + \alpha.$$

Show further that on the positive characteristic through S,

$$\theta - f(\mu) = -2\alpha - f(\mu_1)$$

and infer that θ and μ are both constant on this characteristic.

Writing the boundary as $y = -F(x)$, show that the solution at the point (x, y) in the simple wave region is

$$\theta(x, y) = -\alpha, \qquad f(\mu(x, y)) = f(\mu_1) + \alpha,$$

where $\alpha = \tan^{-1} F'(\xi)$ and $(y + F(\xi))/(x - \xi) = \theta + \mu$.

Now use Figure 5.5 to show that $\mu \to 0$ when α increases to $\pi/2\lambda - f(\mu_1)$ and hence, that the greatest angle through which a flow can be turned is $(\pi/2)(1/\lambda - 1)$.

5.17 Specialize the answer to Exercise 5.16 to the case when $F(x) = x \tan \alpha$ to show that in the simple wave region, θ and μ are given by

$$\tan(\theta + \mu) = \frac{y}{x}, \qquad \theta + f(\mu) = f(\mu_1).$$

How could you show that θ and μ are only functions of y/x without deriving these equations?

Show that the simple wave region is

$$\tan \mu_1 > \frac{y}{x} > \tan(-\alpha + f^{-1}(\alpha + f(\mu_1))).$$

R5.18 Fluid of depth s_0^2/g is contained in a tank $-1 < x < 0$, and at time $t = 0$, the right-hand wall is moved in a positive direction with speed $U(< 2s_0)$ while the left-hand wall is held fixed at $x = -1$. Assuming that the shallow water equations are valid in the subsequent flow, draw a characteristic diagram and show that

$$u = \frac{2}{3}\left(\frac{x}{t} + s_0\right)$$

in a region bounded by $x + s_0 t = 0$, $x + (s_0 - 3U/2)t = 0$, and $x = 2s_0 t - 3(s_0 t)^{1/3}$.

*5.19 Gas at rest with speed of sound c_0 is contained between diaphragms at $x = \pm a$ in a long tube. At $t = 0$, the diaphragms are broken and the gas flows along the tube in both directions into a vacuum. Sketch the characteristics of the subsequent flow in the (x, t) plane. Show that in $x > 0$, there is an expansion fan (simple wave flow) in which the solution is

$$u = \frac{2}{\gamma + 1}\left(\frac{x - a}{t} + c_0\right), \qquad c = \frac{\gamma - 1}{\gamma + 1}\left(\frac{2c_0}{\gamma - 1} - \frac{x - a}{t}\right)$$

and that this region is bounded by the curves

$$x = a - c_0 t, \qquad x = a + \frac{2c_0}{\gamma - 1}t, \qquad x = a + \frac{2c_0}{\gamma - 1}t - 2\nu c_0^{1-1/\nu}a^{1/\nu}t^{1-1/\nu},$$

where $\nu = \frac{\gamma + 1}{2(\gamma - 1)}$.

In order to solve the problem when the expansion fans interact, we need to change to new variables. By writing $2r = u + 2c/(\gamma - 1)$ and $2s = -u + 2c/(\gamma - 1)$, show that the equations of the characteristics (5.10) reduce to

$$\frac{\partial x}{\partial s} = \left(\frac{\gamma + 1}{2}r - \frac{3 - \gamma}{2}s\right)\frac{\partial t}{\partial s}$$

and

$$\frac{\partial x}{\partial t} = \left(\frac{3 - \gamma}{2}r - \frac{(\gamma + 1)}{2}s\right)\frac{\partial t}{\partial r}.$$

Hence, show that the equation for t as a function of r, and s is

$$\frac{\partial^2 t}{\partial r \partial s} + \frac{\nu}{(r + s)}\left(\frac{\partial t}{\partial r} + \frac{\partial t}{\partial s}\right) = 0. \qquad (*)$$

Show that the boundary conditions on

$$x = \pm\left(a + \frac{2c_0 t}{\gamma - 1} - 2\nu c_0 t\left(\frac{a}{c_0 t}\right)^{1/\nu}\right)$$

transform to

$$t = \frac{a}{c_0}\left[\frac{2c_0}{(\gamma - 1)(s + \frac{c_0}{\gamma - 1})}\right]^\nu \quad \text{on } r = \frac{c_0}{\gamma - 1}$$

and

$$t = \frac{a}{c_0}\left[\frac{2c_0}{(\gamma - 1)(r + \frac{c_0}{\gamma - 1})}\right]^\nu \quad \text{on } s = \frac{c_0}{\gamma - 1}.$$

It is possible to solve for $t(r, s)$ explicitly in terms of hypergeometric functions by using the Riemann function technique (see Garabedian [22]).

However, it is amusing to note that if we set $r + s = \xi$, and $r - s = \eta$, equation (*) becomes

$$\frac{\partial^2 t}{\partial \xi^2} + \frac{2\nu}{\xi}\frac{\partial t}{\partial \xi} = \frac{\partial^2 t}{\partial \eta^2},$$

which is just the wave equation in $2\nu + 1$ dimensions if ξ is identified with the radial direction and η is identified with time. Therefore, we can use the general solutions of Exercise 4.23 of Chapter 4 whenever $2\nu + 1$ is an integer. Taking $2\nu + 1 = n$ implies that $\gamma = n/(n-2)$, so for air with $\gamma = 7/5$, we need to solve the wave equation in seven dimensions!

5.20 Show that in unsteady one-dimensional gas flow, the continuity equation

$$\frac{\partial \rho}{\partial t} + \frac{\partial}{\partial x}(\rho u) = 0$$

implies the existence of a function ξ such that

$$\frac{\partial \xi}{\partial x} = \rho, \qquad \frac{\partial \xi}{\partial t} = -\rho u.$$

Deduce that with ξ and t as independent (Lagrangian) variables,

$$\frac{\partial x}{\partial \xi} = \frac{1}{\rho}, \qquad \frac{\partial^2 x}{\partial t^2} = -\frac{\partial p}{\partial \xi}, \quad \text{and} \quad \frac{\partial}{\partial t}\left(\frac{p}{\rho^\gamma}\right) = 0.$$

5.21 Suppose that a shallow layer of water flows down an inclined plane that makes a small angle α with the horizontal. Show that if the x axis is along the line of greatest slope and the y axis is perpendicular to the plane, then two-dimensional shallow water flow is governed by the equations

$$\frac{\partial u}{\partial x} + \frac{\partial v}{\partial y} = 0,$$

$$\frac{\partial u}{\partial t} + u\frac{\partial u}{\partial x} + v\frac{\partial u}{\partial y} = -\frac{1}{\rho}\frac{\partial p}{\partial x} + g\alpha,$$

$$0 = -\frac{1}{\rho}\frac{\partial p}{\partial y} - g,$$

if α is suitably small. If the surface of the water is given by $y = \eta(x, t)$, show that $p = -\rho g(y - \eta)$ and that on $y = \eta$, $v = \partial\eta/\partial t + u(\partial\eta/\partial x)$. Deduce that

$$\frac{\partial \eta}{\partial t} + \frac{\partial}{\partial x}(u\eta) = 0$$

and

$$\frac{\partial u}{\partial t} + u\frac{\partial u}{\partial x} + g\frac{\partial \eta}{\partial x} = g\alpha,$$

and show that in steady flow with $u\eta = m$,

$$\frac{1}{2}\frac{m^2}{\eta^2} + g\eta - g\alpha x = \text{constant}.$$

Show also that there is a possible solution with η constant in which the flow has constant acceleration $g\alpha$ down the plane.

6

Shock Waves

6.1 Discontinuous Solutions

The time has come to face up to the task of making a mathematical model that can deal with flows containing *shock waves* or *shocks*, across which the various dependent physical variables themselves have discontinuities. Such discontinuities are often called *jump discontinuities*, in contrast to situations in which only the derivatives of the physical variables have discontinuities.

We have already been motivated to study such shock waves in our study of both resonance in Section 4.2 and nozzle flow in Section 4.6.2 of Chapter 4. In neither case were we able to find a physically acceptable smooth solution and we were thus led to postulate the possibility of jump discontinuities. Even more compelling, however, was the analysis of Section 5.3 of Chapter 5, where we saw clear evidence that nonlinear wave propagation frequently leads to a breakdown in the continuity of the flow variables. Not only did we find that smooth solutions could fail to exist if discontinuities were imposed in either the boundary or the initial data as, for example, in the impulsively started piston problem of Section 5.3.1, but, more interestingly, we have seen that discontinuities could arise spontaneously in certain flows in which the data are arbitrarily smooth.

Our theory will apply almost exclusively to systems of partial differential equations that can be written in the form

$$\frac{\partial \mathbf{P}}{\partial t} + \sum_i \frac{\partial \mathbf{Q}_i}{\partial x_i} = \mathbf{R},$$

where $\mathbf{P}$, $\mathbf{Q}_i$, and $\mathbf{R}$ are functions only of the dependent variable $\mathbf{u}$ and the independent variables. Such systems are called systems of *conservation laws*, and we have already seen many examples. Indeed, in our basic gasdynamic model, (2.6) of Chapter 2 is already such a law for mass conservation and (2.7) and (2.8) are consequences of conservation laws for momentum and energy.

6.1.1 Introduction to Weak Solutions

The problem of finding a discontinuous solution to a model which is posed as a system of partial differential equations may be clarified by consideration of a paradigm involving just a single variable.

We suppose that in the one-dimensional flow of a "continuum" of density $\rho(x, t)$, the mass flux, ρu, is a *prescribed* function $f(\rho)$. An example of such a situation would be the simplest "continuum" model for traffic flow (Whitham [23]) and the resulting solution is often called a *kinematic wave*, as already encountered in Section 5.1.

For a continuous flow, the equation of conservation of mass (2.6) will be

$$\frac{\partial \rho}{\partial t} + \frac{\partial}{\partial x} f(\rho) = 0, \tag{6.1}$$

and we have seen already in Section 5.1 that such equations can readily admit multivalued solutions. We suppose that the single-valued solution to this problem has a discontinuity at $x = X(t)$, where ρ jumps from ρ_1 to ρ_2 as shown in Figure 6.1.

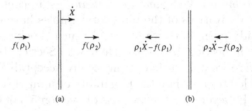

(a)   (b)

Fig. 6.1. (a) A discontinuous solution to (6.1). (b) The fluxes relative to axes moving with the shock.

We can derive an equation for $\dot{X}$ by a simple conservation of mass argument. We first change to axes moving with the shock (Figure 6.1b) and then, relative to the now stationary shock, the mass flux on either side of the shock is $f(\rho_i) - \rho_i \dot{X}$. Thus, the mass flux discrepancy across the shock is

$$[f(\rho) - \rho \dot{X}],$$

where, as is usual, the square brackets denote the size of the jump of the enclosed quantity. Thus, if we prohibit any sources or sinks of mass,

$$[f(\rho)] = [\rho]\dot{X}, \tag{6.2}$$

and this physically derived law will hold across any such shock. We might hope that this *jump condition* (6.2) might be sufficient to determine the position of a shock uniquely, but we will see that this is not necessarily the case.

However, before we study the uniqueness question more closely, we first think more generally about how we might have derived the condition (6.2).

The key idea is to generalize the *derivation* of the equation of conservation of mass, so that no assumption about the differentiability of the dependent variables is needed. For our one-dimensional continuum, the mass in the interval $a(t) < x < b(t)$ is

$$m(t) = \int_{a(t)}^{b(t)} \rho \, dx,$$

where $a(t)$ and $b(t)$ are any functions of t. The mass flux into (a, b) at time t will be

$$q(t) = (f(\rho) - \rho \dot{a})_{x=a(t)} - (f(\rho) - \rho \dot{b})_{x=b(t)},$$

and, hence, in the time interval $(t, t + \delta t)$, the mass balance can be written as

$$\delta m = m(t + \delta t) - m(t) = q(t) \delta t. \tag{6.3}$$

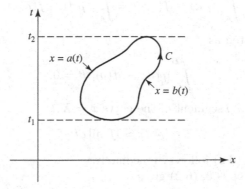

Fig. 6.2. The contour C in the (x, t) plane.

Now, if we consider a closed curve C consisting of the curves $x = a(t)$ and $x = b(t)$ for $t_1 < t < t_2$ in the (x, t) plane as shown in Figure 6.2 and integrate (6.3) from t_1 to t_2, we see that

$$0 = \oint_C f(\rho) \, dt - \rho \, dx, \tag{6.4}$$

where C is *any* closed curve[1] in the plane. By using the divergence theorem, this relation is trivially equivalent to (6.1) for flows in which ρ is a differentiable function. On the other hand, if ρ has a discontinuity on the shock

---

[1] If the interior of C is not a convex region, we may have to divide it up into convex regions in order to do the integration illustrated in Figure 6.3, but the final result (6.4) will still hold.

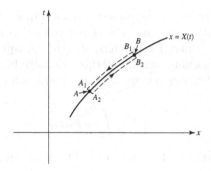

Fig. 6.3. The contour $A_1A_2B_1B_2A_1$ for the discontinuity at $x = X(t)$.

$x = X(t)$, then by taking C to be the contour $A_1A_2B_2B_1A_1$ shown in Figure 6.3, we can see that as $A_1, A_2 \to A$ and $B_1, B_2 \to B$, so that A_1B_1 and A_2B_2 lie along $x = X(t)$ on opposite sides of the shock, (6.4) gives

$$\int_{A_1}^{B_1} \rho \, dx - f(\rho) \, dt = \int_{A_2}^{B_2} \rho \, dx - f(\rho) \, dt.$$

This may be written as

$$\int_{A}^{B} [\rho] \, dx - [f(\rho)] \, dt = 0,$$

and since AB is *any* segment of the curve $x = X(t)$,

$$[\rho] \, dx = [f(\rho)] \, dt$$

on the shock and (6.2) follows immediately.

Note that as $[\rho] \to 0$, (6.2) gives

$$\dot{X} = \lim_{\rho_2 \to \rho_1} \left\{ \frac{f(\rho_2) - f(\rho_1)}{\rho_2 - \rho_1} \right\} = f'(\rho_1),$$

and, so in the limiting case of a weak shock, the shock lies along the characteristic of (6.1). We will find later that this is a general result.

This approach, which uses the integral formulation (6.4) of the problem, means that we can cater for discontinuous as well as continuous solutions and these more general solutions are known as *weak solutions*. For our purposes, a weak solution (this adjective must not be confused with its use in the previous paragraph!) is a solution that is piecewise smooth and which satisfies (6.2) at any points of discontinuity. There is an extensive theory for weak solutions of partial differential equations which are conservation laws, in which the above arguments are made rigorous by the use of *test functions*.[2] However, because

---

[2] The basic idea of a weak solution to (6.1) is to replace (6.4) by $\iint_S (f(\partial\phi/\partial x) + \rho(\partial\phi/\partial t)) \, dx \, dt = 0$, where S is an arbitrary fixed region in the (x, t) plane and ϕ is *any* suitably smooth test function (Ockendon et al. [9]).

the models we consider here are motivated by physical considerations, we can appeal directly to arguments such as those used above.

We now summarize the above ideas before applying them to more general situations. If a scalar conservation law can be written in the integral form

$$\oint_C P\,dx - Q\,dt = 0, \tag{6.5}$$

where P and Q are functions of x and t, and C is any smooth closed curve in the domain in which the solution is sought, then a differentiable solution will satisfy the partial differential equation

$$\frac{\partial P}{\partial t} + \frac{\partial Q}{\partial x} = 0, \tag{6.6}$$

and, across a shock $x = X(t)$,

$$\frac{dx}{dt} = \frac{[Q]}{[P]}. \tag{6.7}$$

We still have to be very careful when using physical arguments that we have the physically acceptable form for the integral formulation (6.5). For example, (6.1) could equally well be written as

$$\frac{\partial}{\partial t}\left(\frac{1}{2}\rho^2\right) + \frac{\partial}{\partial x}F(\rho) = 0,$$

where $F'(\rho) = \rho f'(\rho)$, and this would correspond to an integral formulation

$$\oint_c \frac{1}{2}\rho^2\,dx - F(\rho)\,dt = 0$$

and a jump condition

$$[\tfrac{1}{2}\rho^2]\dot{X}(t) = [F(\rho)],$$

which is quite different from (6.2). Thus, there may be a number of integral formulations corresponding to the same basic differential equation, each of which will give rise to a different jump condition.

Even when we have decided on the correct jump conditions, it may still be possible to find a number of possible discontinuous solutions. For instance, let us consider (6.1) and jump condition (6.2) with $f(\rho) = \frac{1}{2}\rho^2$ and initial conditions

$$\rho(x,0) = \begin{cases} 0, & x < 0 \\ 1, & x > 0. \end{cases}$$

It is easy to see that three possible weak solutions are

$$\rho(x,t) = \begin{cases} 0, & x < \frac{1}{2}t \\ 1, & x > \frac{1}{2}t, \end{cases} \tag{6.8}$$

$$\rho(x,t) = \begin{cases} 0, & x < \frac{1}{4}t \\ \frac{1}{2}, & \frac{1}{4}t < x < \frac{3}{4}t \\ 1, & x > \frac{3}{4}t \end{cases} \tag{6.9}$$

and

$$\rho(x,t) = \begin{cases} 0, & x < 0 \\ \frac{x}{t}, & 0 < x < t \\ 1, & x > t, \end{cases} \qquad (6.10)$$

and we can construct many more. In crude terms, the generalization involved in formulating the problem as (6.5) is dangerous; it allows for discontinuous solutions, but at the moment, it allows far too many of them. Thus, we will need to appeal to some extra information in order to decide which of the many possible weak solutions is relevant to the physical situation.

This problem of non-uniqueness may be resolved in three distinct ways, as discussed in Ockendon et al. [9]. The first method is to use specific physical arguments, the second is to use general thermodynamic principles, and the third is to use the principle of *causality*, which basically asserts that the future cannot influence the past. We will return to the latter two ideas later in the chapter and just employ the first method here.

We suppose that our continuum model (6.1) is generalized to allow for small diffusional effects. Thus, whenever ρ varies spatially, we assume that there will be a small diffusion flux, proportional to $-\partial\rho/\partial x$, which transports material from higher to lower densities. This introduces a second-order term into (6.1), so that in the case when $f = \frac{1}{2}\rho^2$, we are led to consider *Burgers' equation*

$$\frac{\partial \rho}{\partial t} + \rho \frac{\partial \rho}{\partial x} = \varepsilon \frac{\partial^2 \rho}{\partial x^2}, \qquad (6.11)$$

where ε is a small *positive* constant.

If we suppose that a wave $\rho(x,t)$ advances into ambient material $\rho = \rho_1$ as $x \to +\infty$, it is reasonable to seek a traveling wave in which $\rho = \rho(x - Vt)$ and $V > 0$ is constant. It is easy to see that no interesting smooth traveling wave can exist when $\varepsilon = 0$, but for $\varepsilon > 0$, we obtain

$$\varepsilon \rho' = \frac{1}{2}\rho^2 - V\rho + \text{constant},$$

where the prime denotes differentiation with respect to $x - Vt$. Thus, if $\rho \to \rho_2$ as $x - Vt \to -\infty$,

$$\varepsilon \rho' = \frac{1}{2}(\rho - \rho_1)(\rho - \rho_2)$$

and $V = \frac{1}{2}(\rho_1 + \rho_2)$ which, with $V = \dot{X}$, is just jump condition (6.2)! Moreover, by sketching the solutions in the $(\rho, x - Vt)$ plane, we can see that a solution which allows a transition from ρ_1 to ρ_2 when $\varepsilon > 0$ is only possible if $\rho' < 0$ and $\rho_1 \le \rho_2$. The width of the transition region is $O(\varepsilon)$ and the fact that, even as $\varepsilon \to 0$, the density *increases* as the wave passes leads us to the "selection principle" that we need to ensure that there is just one physically acceptable solution satifying the jump condition (6.2). When this principle is applied to the above example, we find that all of the discontinuous solutions such as

(6.8) and (6.9) violate this condition and the only acceptable solution is the continuous "expansion" given by (6.10). Note the importance of the sign of the diffusion coefficient ε and the choice of $f(\rho) = \frac{1}{2}\rho^2$ in this argument; had ε been negative or if f had been a cubic in ρ, say, we would have been led to a different selection principle.

We now have the beginnings of a systematic theory for (6.1) with $f = \frac{1}{2}\rho^2$. This theory eventually leads to the result that, on the whole line $-\infty < x < \infty$, all weak solutions tend, as $t \to \infty$, to a combination of *N-waves*, in which linear segments of the form $(x + x_0)/(t + t_0)$ where x_0 and t_0 are constants, are separated by jump discontinuities satisfying (6.2) (see Lax[24]).

It can be shown (Ockendon and Tayler [1]; the argument is similar to that used in Exercise 5.10) that the introduction of weak viscosity into the equations of gasdynamics leads systematically to (6.11) as the model for the pressure or density in a weak shock wave. Hence, the requirement that the shock is *compressive*, so that the gas pressure increases as the shock passes, is indeed the appropriate selection principle for a physically acceptable gas-dynamic shock.

The idea of weak solutions can be generalized to flows in more than one space dimension. Suppose, for example, that in two dimensions, the conservation law for a differentiable flow is

$$\frac{\partial \rho}{\partial t} + \frac{\partial}{\partial x}f(\rho) + \frac{\partial}{\partial y}g(\rho) = 0; \qquad (6.12)$$

then (6.4) is replaced by

$$\int\int_S (\rho, f, g) \cdot \mathbf{dS} = 0, \qquad (6.13)$$

where S is any smooth closed surface in (t, x, y) space. Writing S as $F(t, x, y) = 0$, so that

$$\mathbf{dS} = \left(\frac{\partial F}{\partial t}, \frac{\partial F}{\partial x}, \frac{\partial F}{\partial y}\right)\left(\left(\frac{\partial F}{\partial t}\right)^2 + \left(\frac{\partial F}{\partial x}\right)^2 + \left(\frac{\partial F}{\partial y}\right)^2\right)^{-1/2} dS,$$

and noting that the normal velocity v_n of a point on S is

$$v_n = -\frac{\partial F/\partial t}{\left((\partial F/\partial x)^2 + (\partial F/\partial y)^2\right)^{1/2}},$$

then an argument analogous to that used to derive (6.2) leads to the jump condition

$$[\rho]v_n = [(f, g) \cdot \mathbf{n}], \qquad (6.14)$$

where $\mathbf{n} = (\partial F/\partial x, \partial F/\partial y)/((\partial F/\partial x)^2 + (\partial F/\partial y)^2)^{1/2}$ is the normal to the projection of S in the x, y plane at time t.

We can remark here that the free surfaces in our gravity wave models in Chapter 3 can be regarded as jumps from $\rho = 0$ on one side of the boundary (the air) to $\rho = $ constant in the water. Hence, corresponding to the continuity equation (2.6), the jump condition derived from (6.14) leads directly to the free-boundary condition (3.10).

6.1.2 Rankine–Hugoniot Shock Conditions

We now consider one-dimensional unsteady gas flow with shocks. Motivated by the previous section, we start by writing down integral formulations for conservation of mass, momentum, and energy.

We can use (6.4) directly to write down conservation of mass in the form

$$\oint_C \rho \, dx - \rho u \, dt = 0, \tag{6.15}$$

and thus, from (6.7), across a shock

$$\dot{X} = \frac{[\rho u]}{[\rho]}. \tag{6.16}$$

In a similar way, the conservation of momentum is written as

$$\oint_C \rho u \, dx - (\rho u^2 + p) \, dt = 0, \tag{6.17}$$

to give the shock condition

$$\dot{X} = \frac{[\rho u^2 + p]}{[\rho u]}. \tag{6.18}$$

Finally, the energy equation is

$$\oint_C \left(\frac{1}{2}\rho u^2 + \rho e\right) dx - \left(\frac{1}{2}\rho u^3 + \rho e u + p u\right) dt = 0, \tag{6.19}$$

and, on putting $e = c_v T = p/(\gamma - 1)\rho$, this leads to the third shock condition

$$\dot{X} = \frac{[\frac{1}{2}\rho u^3 + \gamma p u/(\gamma - 1)]}{[\frac{1}{2}\rho u^2 + p/(\gamma - 1)]}. \tag{6.20}$$

Note that in the same way that we were able to go from (6.5) to (6.6), we can easily derive the equations for one-dimensional flow without shocks from the integral formulations (6.15), (6.17), and (6.19) and, after some manipulation, we arrive at (5.3)–(5.5) as expected. However, we emphasize that the shock relation (6.20) must be derived from the *conservation* form of the energy equation and we note that in spite of (5.5), p/ρ^γ will *not* be conserved across a shock.

The three shock relations or jump conditions given by (6.16), (6.18), and (6.20) are called the *Rankine–Hugoniot relations* and are usually written in the form

$$[\rho(u - \dot{X})] = 0, \tag{6.21}$$

$$[p + \rho(u - \dot{X})^2] = 0 \tag{6.22}$$

and

$$\left[\frac{\gamma p}{(\gamma - 1)\rho} + \frac{1}{2}(u - \dot{X})^2\right] = 0. \tag{6.23}$$

Equation (6.21) comes directly from (6.16), but some algebraic manipulation is needed to obtain (6.22) and (6.23), (Exercise 6.1).

It can be seen from (6.21)–(6.23) that it is the velocity *relative* to the shock that appears naturally in the shock relations. This is not surprising since, as we saw in Section 6.1.1, physically motivated jump conditions are most conveniently written down by considering the flow *relative to the shock*.

Further properties of shocks can be understood by rewriting jump conditions in terms of M_1, the upstream Mach number relative to the shock. We use the suffix 1 for upstream variables ahead of the shock and the suffix 2 for downstream variables behind the shock and put $M_i = (\dot{X} - u_i)/c_i$. Then, as shown in Exercise 6.2, the downstream variables may be expressed in terms of the upstream ones as

$$\frac{p_2}{p_1} = \frac{2\gamma M_1^2}{\gamma + 1} - \frac{\gamma - 1}{\gamma + 1}, \tag{6.24}$$

$$\frac{\rho_2}{\rho_1} = \frac{\dot{X} - u_1}{\dot{X} - u_2} = \frac{(\gamma + 1)M_1^2}{2 + (\gamma - 1)M_1^2} \tag{6.25}$$

and

$$M_2^2 = \frac{(\gamma - 1)M_1^2 + 2}{2\gamma M_1^2 - (\gamma - 1)}. \tag{6.26}$$

Now, following the clue from the end of Section 6.1.1, we demand that the shock be *compressive* so that

$$p_2 \geq p_1. \tag{6.27}$$

Observations reveal that this inequality is satisfied in practice by all purely gasdynamic shock waves and it implies from (6.24)–(6.26) that

$$M_1 \geq 1, \qquad M_2 \leq 1, \quad \text{and} \quad \rho_2 \geq \rho_1. \tag{6.28}$$

Thus, relative to the shock, the flow upstream (ahead) must be supersonic and the flow downstream (behind) is subsonic. A less immediate implication of (6.27) is that

$$\frac{p_2}{\rho_2^\gamma} \geq \frac{p_1}{\rho_1^\gamma}, \tag{6.29}$$

which shows that the entropy of the gas always *increases* as it passes through the shock (Exercise 6.3). Indeed, were we to be guided by thermodynamics, we could appeal to the Second Law to assert (6.29) and then deduce (6.27). The equivalence of this approach to the one we have adopted is unsurprising since the dissipation inherent in asserting that $dQ/dt \geq 0$ in (2.16) is also inherent in asserting that the coefficient of viscosity of the gas, which is analogous to the diffusion coefficient ε in (6.11), is positive. Note also that the increase in entropy across a shock depends on the shock speed $\dot{X}$ and this means that we can never have homentropic flow downstream of a genuinely unsteady shock for which $\ddot{X} \neq 0$.

We now illlustrate these results with an example. We reconsider the piston problem discussed in Section 5.3.1, but now we suppose that the piston is pushed *into* the gas in $x > 0$ with constant speed V. Let us look for a solution in which a compressive shock travels into the quiescent gas with constant speed $\dot{X}$ so that, behind this shock, the gas moves with the piston at constant speed V. We assume the pressure and density ahead of the shock are p_0 and ρ_0, respectively, and behind the shock are p_1 and ρ_1, respectively. Then, writing down the three Rankine–Hugoniot equations allows us to determine the three unknown quantities $\dot{X}$, p_1, and ρ_1. Eliminating p_1 and ρ_1, we obtain the equation

$$\dot{X}^2 - \frac{(\gamma+1)}{2}V\dot{X} - c_0^2 = 0 \tag{6.30}$$

for $\dot{X}$, where $c_0^2 = \gamma p_0/\rho_0$. Since $\dot{X}$ must be positive,

$$\dot{X} = \frac{\gamma+1}{4}V + \left[\frac{(\gamma+1)^2 V^2}{16} + c_0^2\right]^{1/2},$$

and the characteristics are as shown in Figure 6.4. The slope of the shock is *between* the slopes of the upstream and downstream positive characteristics and this configuration, which is typical of all evolutionary shocks, can be shown to be a manifestation of the principle of causality, which is described in detail in Ockendon et al. [9]. In the limiting case, as $V \to 0$ and the shock becomes very weak, the shock can again be seen to lie along the positive characteristic $x = c_0 t$.

6.1.3 Shocks in Two-dimensional Steady Flow

It is quite easy to generalize our results for a one-dimensional steady shock with $\dot{X} = 0$ to a straight oblique shock simply by imposing a velocity parallel to the shock on the whole system. However, we proceed more systematically by deriving the shock relations *ab initio* by using the method of weak solutions for two-dimensional steady flow.

The integral equations of motion in this case are even easier to write down than those of the previous section. Conservation of mass of a fluid in steady

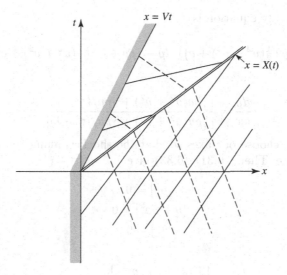

Fig. 6.4. The compressive piston problem: $-\,-\,-\,-$ positive characteristics, $-\,-\,-$ negative characteristics.

flow implies that

$$\oint_C \rho \mathbf{u} \cdot \mathbf{n}\, dS = 0$$

around any contour C, and so

$$\oint_C \rho u\, dy - \rho v\, dx = 0.$$

Hence, from (6.7), the jump condition is

$$\frac{dy}{dx} = \frac{[\rho v]}{[\rho u]}. \tag{6.31}$$

There are two components of the momentum equation, and in the x and y directions, we obtain

$$\oint_C (p + \rho u^2)\, dy - \rho uv\, dx = 0$$

and

$$\oint_C \rho uv\, dy - (p + \rho v^2)\, dx = 0.$$

Thus, the shock relations for the momentum are

$$\frac{dy}{dx} = \frac{[\rho uv]}{[p + \rho u^2]} = \frac{[p + \rho v^2]}{[\rho uv]}. \tag{6.32}$$

Finally, the energy equation is

$$\oint_C \left(pu + \rho u \left(\tfrac{1}{2}(u^2 + v^2) + e\right)\right) dy - \left(pv + \rho v \left(\tfrac{1}{2}(u^2 + v^2) + e\right)\right) dx = 0,$$

so that

$$\frac{dy}{dx} = \frac{\left[\tfrac{1}{2}\rho v(u^2 + v^2) + \gamma pv/(\gamma - 1)\right]}{\left[\tfrac{1}{2}\rho u(u^2 + v^2) + \gamma pu/(\gamma - 1)\right]}. \tag{6.33}$$

Let us now choose our axes so that the shock is along the y axis and so dy/dx is infinite. Then, (6.31)–(6.33) give

$$[\rho u] = 0, \tag{6.34}$$

$$[p + \rho u^2] = 0, \tag{6.35}$$

$$[v] = 0 \tag{6.36}$$

and

$$\left[\frac{1}{2}u^2 + \frac{\gamma p}{(\gamma - 1)\rho}\right] = 0, \tag{6.37}$$

which are, as expected, the Rankine–Hugoniot equations normal to a steady shock as derived in Section 6.1.2, with the extra condition that the velocity parallel to the shock is conserved.

Since the condition (6.27) that the shock be compressive implies that the velocity of the gas normal to the shock is decreased as it passes through a shock, we can see immediately from Figure 6.5 that the effect of the shock is to turn the flow *toward* the shock. Note also that it is only the component of the velocity normal to the shock that must be supersonic ahead of the shock and subsonic behind it. Thus, although the overall flow must be supersonic ahead of the shock, so that $M_1 > 1$, we only know that $M_2 \sin(\beta - \theta) < 1$. Hence, the flow downstream could be either subsonic or supersonic. Indeed, transonic aeroplanes have swept back wings partly to lessen the possibility of shocks forming at the leading edges of the wings.

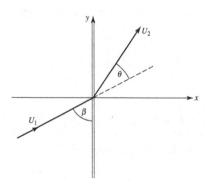

Fig. 6.5. Flow deflection due to an oblique shock.

We remark that (6.31)–(6.33) are all direct consequences of the unsteady shock relation (6.14). We may also expect the relations (6.34)–(6.37) to hold across any smoothly evolving shock, straight or curved, as long as the jumps are taken with respect to axes moving with the shock and aligned so that the x axis is in the direction of the normal to the shock.

It is often convenient to work with the actual velocities before and after a stationary oblique shock rather than with the components along and perpendicular to the shock. We therefore rewrite (6.34)–(6.37) in terms of U_1 and U_2 as indicated in Figure 6.5 to get

$$\rho_1 U_1 \sin\beta = \rho_2 U_2 \sin(\beta - \theta), \tag{6.38}$$

$$p_1 + \rho_1 U_1^2 \sin^2\beta = p_2 + \rho_2 U_2^2 \sin^2(\beta - \theta), \tag{6.39}$$

$$U_1 \cos\beta = U_2 \cos(\beta - \theta) \tag{6.40}$$

and

$$\frac{1}{2}U_1^2 \sin^2\beta + \frac{\gamma p_1}{(\gamma - 1)\rho_1} = \frac{1}{2}U_2^2 \sin^2(\beta - \theta) + \frac{\gamma p_2}{(\gamma - 1)\rho_2}. \tag{6.41}$$

We note that using (6.40), (6.41) can be written as

$$\frac{1}{2}U_1^2 + \frac{\gamma p_1}{(\gamma - 1)\rho_1} = \frac{1}{2}U_2^2 + \frac{\gamma p_2}{(\gamma - 1)\rho_2}. \tag{6.42}$$

Now, we can once again manipulate these equations in terms of M_1 and β so that instead of (6.24)–(6.26), we have

$$\frac{p_2}{p_1} = \frac{2\gamma M_1^2 \sin^2\beta}{\gamma + 1} - \frac{\gamma - 1}{\gamma + 1}, \tag{6.43}$$

$$\frac{\rho_2}{\rho_1} = \frac{\tan\beta}{\tan(\beta - \theta)} = \frac{(\gamma + 1)M_1^2 \sin^2\beta}{2 + (\gamma - 1)M_1^2 \sin^2\beta} \tag{6.44}$$

and

$$M_2^2 \sin^2(\beta - \theta) = \frac{(\gamma - 1)M_1^2 \sin^2\beta + 2}{2\gamma M_1^2 \sin^2\beta - (\gamma - 1)}. \tag{6.45}$$

Although these shock relations, together with the condition of compression or entropy increase across the shock, appear straightforward enough, alarming complexities can arise even in the simplest application. Consider, for example, supersonic flow into a concave corner, as illustrated in Figure 6.6 in which the flow is to be turned through an angle θ. This is analogous to the ingoing piston problem in one-dimensional unsteady flow and we now try to find a straight shock in the (x, y) plane which will turn the flow through the angle θ. Using (6.44), we can deduce that the shock angle β is related to M_1 and θ via the formula

$$\tan(\beta - \theta) = \frac{2 + (\gamma - 1)M_1^2 \sin^2\beta}{(\gamma + 1)M_1^2 \sin\beta\cos\beta}. \tag{6.46}$$

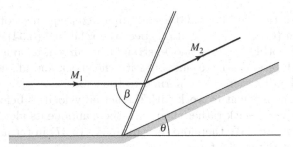

Fig. 6.6. Supersonic flow past a concave corner.

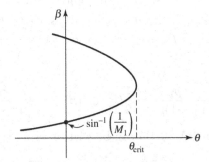

Fig. 6.7. The shock angle β in terms of the deflection θ.

Note that as $\theta \to 0$, $\beta \to \sin^{-1}(1/M_1)$ and the shock is again a characteristic. However, when we plot β against θ in Figure 6.7, we find that there are two possible values for β for each $\theta < \theta_{\mathrm{crit}}$. Now, it can be argued (but the details are beyond the scope of this book) that the upper branch, which represents the stronger shock, is unstable, but in order to decide what happens if $\theta > \theta_{\mathrm{crit}}$, we need some experimental evidence. It turns out that the shock is no longer straight when $\theta > \theta_{\mathrm{crit}}$ and that it also "stands off" from the wedge, as shown in Figure 6.8.[3] Now, we have lost the simple situation of two uniform flows, each with constant entropy, and since behind the curved shock the flow is no longer homentropic, the full equations of gasdynamics will need to be solved there. Note that the shock will be normal to the incoming flow at A and thus the flow immediately behind the shock is always subsonic.

These ideas may be used to understand the supersonic flow past a two-dimensional wing with a sharp leading edge (Figs. 6.9a and 6.9b) and this leads us ultimately to the even more complicated problem of supersonic flow past a *blunt body*, as shown in Figure 6.9c.

---

[3] In view of the dimensionality arguments of Section 5.3.1 and Exercise 5.15, this phenomenon should not occur for a corner in an infinite wall; in practice, the stand-off distance and shock curvature will be determined by the downstream geometry.

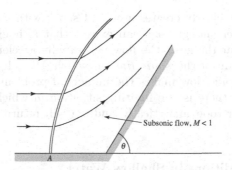

Fig. 6.8. Supersonic flow past a concave corner when $\theta > \theta_{\mathrm{crit}}$.

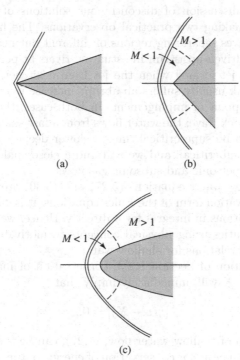

(a)   (b)

(c)

Fig. 6.9. (a) Supersonic flow past a slender wedge $\theta < \theta_{\mathrm{crit}}$. (b) Supersonic flow past a wedge with $\theta > \theta_{\mathrm{crit}}$; ..... indicates the sonic line. (c) Supersonic flow past a blunt body; ..... indicates the sonic line.

The blunt body clearly creates a curved shock with stand off, and if the body is also slender enough, one can expect that although there will be a subsonic region near the nose, the flow will rapidly accelerate to become supersonic downstream of the *sonic line* on which $M = 1$. This change from subsonic to supersonic flow makes the numerical problem particularly troublesome. However, there is one limiting situation in which such blunt body flows can be understood analytically and we will return to this in Section 6.3.4.

6.1.4 Jump Conditions in Shallow Water

We motivate our discussion of discontinuous solutions of the shallow water equations by making two practical observations. The first is that *bores*, which are steep waves separating regions of different but nearly constant water depth, can be driven upstream in estuarine rivers at periods of high tide. Second, the radial jet formed when the jet from a kitchen tap impinges on a flat-bottomed sink usually suffers an abrupt increase in depth at a definite distance from the point of impingement. In both cases, the sudden jump is reminiscent of a shock wave; the water flows from a fast shallow region, which can be observed to be supercritical, into a slower deeper region, which can be observed to be subcritical, and we will find a close analogy between such water flows and supersonic and subsonic gas flows.

Since the shallow water equations (5.27) and (5.30) are only approximations to the conservation form of the Euler equations, it is dangerous to try to formulate the equations in integral form directly. Hence, we study these "hydraulic" discontinuities using physical concepts to motivate the appropriate Rankine–Hugoniot relations for shallow water.

Using the notation of Section 5.2.3, conservation of mass across a bore moving with speed $\dot{X}$ will immediately imply that

$$[\eta(u - \dot{X})] = 0. \tag{6.47}$$

The other equation of shallow water flow, (5.27), can be thought of as either conservation of momentum or conservation of energy, but in a continuous flow, both these quantities are conserved by this one equation, as was the case for the inviscid incompressible Euler equation (2.7). Now, because we only have two first-order equations (5.27) and (5.30), there can only be two Rankine–Hugoniot conditions and hence, across a bore, it is possible to conserve *either* momentum *or* energy but not both. In most situations, ranging from the kitchen sink example to a bore on a river, it is more realistic to conserve momentum, and energy will be dissipated. For weak bores, this energy is mainly transported away from the discontinuity by a train of waves and this is known as an *undular bore*, whereas for a stronger *turbulent bore*, the energy is dissipated in the form of turbulence at a wave crest.[4] Either type of bore may also be referred to as a *hydraulic jump*, although this term is more usually reserved for a stationary discontinuity.

It can also happen that momentum can be destroyed by a sluice gate or a small obstacle on the bed of the river. In this case, a smooth transition can be created for which it may be more appropriate to conserve energy rather than momentum. However, for the rest of this chapter, we will study momentum-conserving bores.

First, we make use of the fact that the change in momentum of the flow on either side of the bore must be balanced by the pressure forces acting. Thus,

$$\left[\rho\eta(u - \dot{X})u\right] + \left[\int_0^\eta (p - p_0)\,dz\right] = 0,$$

and, using (6.47) and remembering that $p - p_0 = \rho g(\eta - z)$, we obtain

$$\left[\rho\eta(u - \dot{X})^2 + \tfrac{1}{2}\rho g\eta^2\right] = 0. \qquad (6.48)$$

Using the now familiar conservation of mass argument and rewriting (6.47) and (6.48) in terms of s, where $s^2 = g\eta$, we obtain the shock relations in the form

$$[s^2(u - \dot{X})] = 0 \qquad (6.49)$$

and

$$\left[s^2(u - \dot{X})^2 + \frac{1}{2}s^4\right] = 0. \qquad (6.50)$$

Note that although the partial differential equations for unsteady one-dimensional gasdynamics can be translated into those for shallow water by putting $c = s$ and $\gamma = 2$, the Rankine–Hugoniot shock relations (6.21)–(6.23) are nothing like (6.49) and (6.50). Hence, although hydraulic tanks can be used in the laboratory to simulate continuous homentropic gas flow, they cannot be used to study gas flows with shocks. However, we remark that we could have obtained (6.49) and (6.50) from the gasdynamic shock conditions (6.21) and (6.22) for a gas with $\gamma = 2$ had we assumed that the entropy, and hence p/ρ^2, was conserved across the shock. We know that entropy is not conserved across a shock in a real gas but a good model for internal shocks in water (not hydraulic jumps) is to use the Rankine–Hugoniot conditions (6.21) and (6.22) together with conservation of entropy $[p/\rho^\gamma] = 0$, with $\gamma \simeq 7$ (Glass and Sislan [4]).

We cannot ignore energy altogether and, indeed, energy considerations are vital when it comes to selecting physically acceptable solutions of (6.49) and (6.50). We must make sure that energy is dissipated rather than gained

---

[4] Practical observations indicate that the transition from an undular bore to a turbulent bore occurs when the ratio of the increase in depth to the original depth is around 0.3.

across a momentum-preserving discontinuity, and when we do this, we derive
a condition which is analogous to that derived from the compressive condition
(6.27) for a gasdynamic shock. As a discontinuity converts fluid from velocity
u_1 and depth η_1 to velocity u_2 and depth η_2, the rate of increase in kinetic
and potential energy is

$$\Delta \dot{E}_1 = \left[\frac{1}{2}\rho\eta u^2(\dot{X} - u) + \int_0^\eta \rho g z(\dot{X} - u)\,dz \right]_1^2$$
$$= \frac{\rho}{2g}[(u^2 + s^2)s^2(\dot{X} - u)]_1^2.$$

However, energy is also created by the work done by the pressure forces, and
the rate at which this work is done is

$$\Delta \dot{E}_2 = \left[\int_0^\eta (p - p_0)u\,dz \right]_2^1$$
$$= -\frac{\rho}{2g}[s^4 u]_1^2.$$

Thus, on using (6.49) and (6.50), the total rate at which energy is gained is

$$\Delta \dot{E}_1 + \Delta \dot{E}_2 = \frac{\rho}{g}s_1^2(\dot{X} - u_1)\left[\frac{1}{2}(u - \dot{X})^2 + s^2 \right]_1^2, \tag{6.51}$$

which can be rewritten as

$$-\frac{\rho(\dot{X} - u_1)(s_2^2 - s_1^2)^3}{4gs_2^2} \tag{6.52}$$

after further manipulation. Thus, we see immediately that energy cannot be
conserved. Moreover, since energy must be lost, we see that if $\dot{X} > u_1$, then
$s_2 > s_1$. Hence, for a momentum-conserving turbulent bore on shallow water,
the flow in front of the discontinuity will be shallower than the flow behind
it. In the same way, in a stationary hydraulic jump, the flow can only jump
from fast shallow flow to a slower deeper flow. It can be shown (Exercise 6.11)
that relative to the bore, the flow ahead of the discontinuity is supercritical
and the flow behind is subcritical; this is exactly analogous to the result for a
plane shock in a gas where, relative to the shock, the flow ahead is supersonic
and the flow behind is subsonic.

  As an example, we consider a piston being pushed with constant velocity
V into static water of depth s_1^2/g. If we assume that a bore runs ahead of
the piston with constant speed $\dot{X}$ and that the depth of the water behind the
bore is s_2^2/g, we can use (6.49) and (6.50) to get

$$s_2^2(V - \dot{X}) = -s_1^2\dot{X}$$

and

$$s_2^2(V - \dot{X})^2 + \tfrac{1}{2}s_2^4 = s_1^2\dot{X}^2 + \tfrac{1}{2}s_1^4.$$

Hence, eliminating s_2, the equation for $\dot{X}$ is

$$\dot{X}^3 - 2V\dot{X}^2 + \dot{X}(V^2 - s_1^2) + \tfrac{1}{2}s_1^2 V = 0,$$

which is very different from (6.30), the equation for the shock speed in the analogous gasdynamic problem.

6.2 Other Flows involving Shock Waves

In this section, we consider a number of more complicated flows in which shocks can arise. First, we consider some gasdynamic examples and then one from shallow water theory, all of the examples being extensions of cases that have been considered earlier in this chapter.

6.2.1 Shock Tubes

We are now in a position to generalize the instantaneously removed piston problem of Section 5.3.1 to the more physically realistic case of the sudden rupture of a diaphragm separating high- and low-pressure gas in a *shock tube*. Immediately after the rupture, the low-pressure gas will be compressed by the high-pressure gas and, at first sight, we are led to conjecture the scenario sketched in Figure 6.10, where a shock propagates into the low-pressure gas on the right and an expansion wave propagates into the high-pressure gas on the left. Thus, we can use the solution in Section 5.3.1 for the expansion wave and the solution from Section 6.1.2 for the shock wave generated by an ingoing piston. However, when we try to piece together these two solutions, we find that it is not possible to construct a solution which is continuous away from the shock. We need to introduce a new kind of discontinuity at the position where the low-pressure gas which has traversed the shock meets the expanded high-pressure gas from the left of the diaphragm. In general, these two regions of gas will have different entropies and so it will be impossible to make both the pressure and the density continuous at this point. If we go back to the Rankine–Hugoniot shock conditions (6.16), (6.18), and (6.20), we can see that it is possible to have discontinuities in which

$$[u] = 0, \qquad [p] = 0, \qquad [\rho] \neq 0 \qquad\qquad (6.53)$$

as long as $u = \dot{X}$ on both sides of the discontinuity. This special solution of the Rankine–Hugoniot equations is called a *contact discontinuity*, and in this problem, a contact discontinuity will travel with the gas particles which were originally adjacent to the diaphragm. Hence, the final scenario illustrated in Figure 6.10 must contain a contact discontinuity along the line OC, as shown in detail in Exercise 6.17. Note that in the limiting case where there is a vacuum in the right-hand half of the tube initially, we retrieve the solution mentioned in Section 5.3.1 when the piston is removed instantaneously.

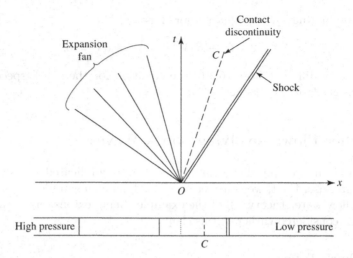

Fig. 6.10. A shock tube.

From a mathematical point of view, it is always helpful to think of jump discontinuities in the limit as they become vanishingly weak. We have already seen in Section 6.1.2 that as the jumps tend to zero, the shock tends to either the positive or negative characteristic. However, when we take the same limit for a contact discontinuity, its path remains the particle path $dX/dt = u$, which is the third characteristic of the system of equations (5.3)–(5.5).

A similar analysis can be performed for a sudden dam break in which there is originally water at different levels on both sides of the dam. A bore will travel into the shallower water and an expansion wave will travel into the deeper water, but in this case, there will be *no* contact discontinuity since the original system of (5.27) and (5.30) has just two characteristics (Exercise 6.14).

6.2.2 Oblique Shock Interactions

We can use the Rankine–Hugoniot conditions for an oblique shock to construct another composite flow, this time in two dimensions. Suppose we consider the flow generated by two corners, as shown in Figure 6.11, and ask what happens downstream of the shock interaction.

Since, from (5.21), we know that the characteristics make angles $\pm \sin^{-1}\left(\frac{1}{M}\right)$ with the flow direction and that the shocks become characteristic when they are weak, it is tempting to postulate, at least for small flow deflections, the existence of two crossing "transmitted" shocks, as shown in Figure 6.12. These shocks must be arranged so that the ultimate flow deflection is the same for all incoming fluid particles. This is indeed possible for certain parameter regimes of the variables M_1, β_1, and β_2, but it is only possible at the expense of admitting a "two-dimensional" contact discontinuity along OP, as shown in

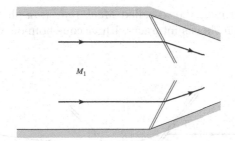

Fig. 6.11. Flow generated by two concave corners.

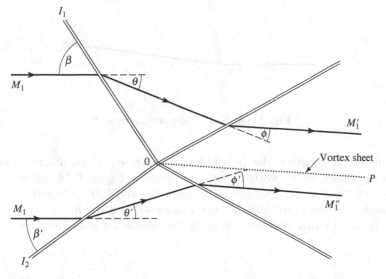

Fig. 6.12. Shock interaction; case 1: incident shocks I_1 and I_2 are at angle β_1 and β_2 to the free stream.

Figure 6.12 (see Exercise 6.18 for details). Note that the velocity perpendicular to the contact discontinuity must be zero on both sides and the jump in p must also be zero, but that from (6.31)–(6.33), the discontinuity can allow both

$$[\rho] \neq 0 \quad \text{and} \quad [v] \neq 0,$$

where v is the component of velocity parallel to the discontinuity. Thus this discontinuity is a *vortex sheet* and we recall from Section 4.4 that a vortex sheet is always subject to the Kelvin–Helmholtz instability. Hence, we expect it to rapidly "smear out" into a turbulent layer that mixes the gas on either side of the sheet.

However, this is not the end of the story because it is sometimes possible for two "incoming" shocks to generate just one outgoing shock and a contact discontinuity as shown in Figure 6.13. When the lower shock is normal to the oncoming stream, this is the so-called *Mach reflection* phenomenon, which is often observed when shocks interact with viscous boundary layers (Liepmann and Roshko [24]).

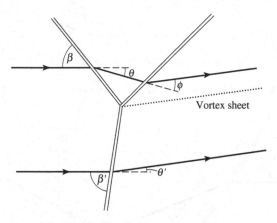

Fig. 6.13. Shock interaction; case 2.

Finally, we mention the possibility of achieving shock attenuation as a result of impact with a vortex sheet, as shown in Figure 6.14. As detailed in Exercise 6.20, the strength of the transmitted shock T is less than that of the incident shock I, and this has been proposed as a method of ameliorating sonic boom. In this context, we note that annoying oblique "bow shocks"

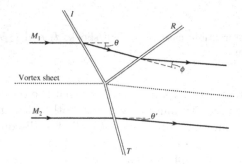

Fig. 6.14. Shock reflection and transmission from a vortex sheet.

are inevitably generated by supersonic aircraft with the one exception of the *Busemann biplane*, which generates a finite system of shocks, as shown in Figure 6.15, and is described in Liepmann and Roshko [25]. Alas, this biplane is only of academic interest since it generates no lift!

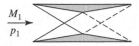

Fig. 6.15. Shock waves generated by the Busemann Biplane.

6.2.3 Steady Quasi-one-dimensional Gas Flow

In Section 4.6.2 we showed how, by using a control-volume argument, the equations of gasdynamics can be reduced to three algebraic equations (4.43)–(4.45) in the case of flow in a nozzle of slowly varying cross-section $A(x)$. Such flows are called *quasi-one-dimensional* and our knowledge of shock waves now enables us to understand the flow through a converging–diverging nozzle more fully.

We first use (4.43)–(4.45) to write A and the pressure p in terms of the Mach number $M = u/c$ as

$$A = \frac{m}{\rho_0 c_0 M} \left\{ 1 + \frac{\gamma - 1}{2} M^2 \right\}^{(\gamma+1)/2(\gamma-1)} \tag{6.54}$$

and

$$p = p_0 \left\{ 1 + \frac{\gamma - 1}{2} M^2 \right\}^{-\gamma/(\gamma-1)}. \tag{6.55}$$

Here, p_0, ρ_0, and c_0 are the pressure, density, and speed of sound, respectively, in a large reservoir which feeds the nozzle. Plotting A against M as in Figure 6.16, we see that the minimum value A_c of this function can only be attained when M is unity and the flow is sonic.[5] Hence, we are led to consider the flow

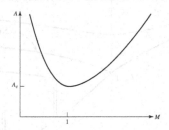

Fig. 6.16. Variation of A with M from (6.54).

from a high-pressure reservoir into a converging–diverging nozzle, known as a *Laval nozzle*, as shown in Figure 6.17.

If the minimum cross-sectional area of the nozzle, $A_{\min}$, is greater than A_c, and the flow entering the nozzle is subsonic, then the flow will always be

---

[5] Note that we have implicitly assumed that the flow is compressible. Clearly, an incompressible flow can flow smoothly through any slowly varying nozzle.

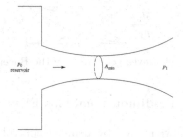

Fig. 6.17. A Laval nozzle.

subsonic, whereas if we can arrange for $A_{\min}$ to equal A_c, it might be possible for the flow to become supersonic in the divergent part of the nozzle. If this can be done, it leads to a design for a supersonic wind tunnel. The flow rate m will be controlled by the downstream pressure p_1 imposed at the end of the nozzle, and by decreasing p_1 from p_0, we will arrive at an exit pressure p_c for which $A_{\min} = A_c$. The question then arises as to "what happens when $p_1 < p_c$?"

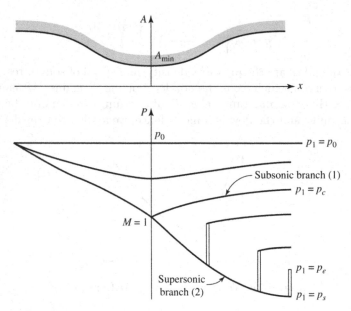

Fig. 6.18. Pressure variation in a nozzle for varying values of the downstream pressure p_1.

In Figure 6.18, we show how the pressure along the nozzle varies for different values of p_1. We can see that if p_1 is between p_0 and p_c, then the flow will remain subsonic throughout. When $p_1 = p_c$, the Mach number will be unity at the throat, although the flow will revert to being subsonic in $x > 0$, as indicated by branch (1) in Figure 6.18. However, the key observation is that there is *another* possible smooth solution with $M = 1$ at the throat; this will correspond to branch (2) with supersonic flow in $x > 0$ and it occurs when $p_1 = p_s$, which can be found from (6.54) and (6.55). When $p_1 < p_c$, the flow is said to be *choked*, and the new phenomenon in such flows is the presence of a shock wave in the supersonic flow downstream of the throat; the flow upstream of the throat is the same for all $p_1 < p_c$. As p_1 decreases from p_c, this shock moves to the right and eventually reaches the end of the nozzle when p_1 attains yet another critical value p_e. For values of p_1 between p_s and p_e, the shock wave is ejected into the downstream atmosphere in a complicated three-dimensional flow involving multiple shock waves. Furthermore, if $p_1 < p_s$, the flow downstream will contain a series of Prandtl–Meyer expansion fans as discussed in Chapman [26].

6.2.4 Shock Waves with Chemical Reactions

The violence inflicted on gas particles as they pass through a shock wave can frequently induce chemical reactions, the most awesome of which is when a combustible gas undergoes an intense exothermic reaction as it encounters the temperature, pressure, and density rise at the shock. Such a configuration is called a *detonation* and here we mention the simplest model for such detonations. We simply sweep all of the chemistry aside and assert that although the mass and momentum conservation laws still apply at the shock, energy is gained at a prescribed rate E per unit mass. Thus, (6.23) becomes

$$\left[\frac{\gamma p}{(\gamma - 1)\rho} + \frac{1}{2}(\dot{X} - u)^2 \right] = E. \tag{6.56}$$

This makes our mathematical analysis, which is complicated enough when $E = 0$, even more difficult to present lucidly. However, great insight can be obtained by noticing that, from (6.24) and (6.25), when $E = 0$, the ratios $\tilde{p} = p_2/p_1$ and $\tilde{\rho} = \rho_2/\rho_1$ satisfy

$$\tilde{p} = \frac{(\gamma + 1) - (\gamma - 1)/\tilde{\rho}}{(\gamma + 1)/\tilde{\rho} - (\gamma - 1)}, \tag{6.57}$$

where the suffix 1 is upstream in the unreacted gas. Thus, the point $(\tilde{p}, \tilde{\rho}^{-1})$ lies on a hyperbola in the $(\tilde{p}, \tilde{\rho}^{-1})$ plane called the *Chapman–Jouguet* curve, as shown in Figure 6.19. The point $(1, 1)$ corresponds to the upstream condition, and the unique compressive solution satisfying the Rankine–Hugoniot condition lies on the part of the hyperbola indicated by the solid line.

Now, let us reintroduce E. Equation (6.57) becomes

$$\tilde{p} = \frac{\gamma + 1 - (\gamma - 1)/\tilde{\rho} + 2\gamma(\gamma - 1)E/c_1^2}{(\gamma + 1)/\tilde{\rho} - (\gamma - 1)}, \tag{6.58}$$

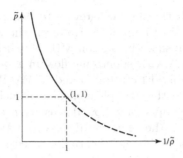

Fig. 6.19. The Chapman-Jouguet curve.

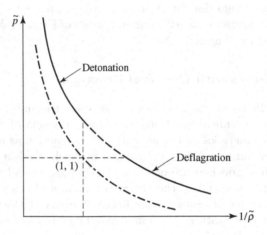

Fig. 6.20. The Chapman-Jouguet curve with constant energy addition.

so that the Chapman–Jouguet curve is displaced as shown in Figure 6.20. The point $(1, 1)$ still represents the upstream condition, but our selection criteria that led us to reject the dashed segment of the hyperbola in Figure 6.19 is no longer available; even the simplest entropy argument would be difficult in the presence of the reaction. It is easy to see that the dashed segment where $\tilde{p} > 1$ and $\tilde{\rho} < 1$ cannot satisfy the Rankine–Hugoniot conditions, but the lower branch where $\tilde{p} < 1$ and $\tilde{\rho} < 1$ cannot be ruled out. This branch represents a *deflagration* and it can be observed in certain circumstances. Further details about this theory can be found in Courant and Friedrichs [27].

6.2.5 Open Channel Flow

Our last example concerns the steady flow of shallow water along a straight channel with a horizontal bottom, but having slowly varying width. This problem is very similar to the gas flow studied in Section 6.2.3 if we make the usual shallow water assumptions that pressure is hydrostatic and that the flow is quasi-one-dimensional. If the width of the channel is $b(x)$, the depth of the water is η, and the horizontal velocity is u, conservation of mass in a control volume gives

$$ub\eta = q, \tag{6.59}$$

where q is the constant flux along the channel. In addition, Bernoulli's equation on the surface streamline leads to

$$\tfrac{1}{2}u^2 + g\eta = gH, \tag{6.60}$$

where H is the pressure head which is the depth of the water in a large reservoir which feeds the channel. We write these equations in terms of the Froude number

$$F = \frac{u}{\sqrt{g\eta}} \tag{6.61}$$

to get

$$\eta = \frac{H}{1 + \tfrac{1}{2}F^2},$$

$$u = F\left(\frac{gH}{1 + \tfrac{1}{2}F^2}\right)^{1/2}$$

and

$$b = \frac{q}{g^{1/2}H^{3/2}F}\left(1 + \frac{1}{2}F^2\right)^{3/2}.$$

Plotting b as a function of F in Figure 6.21 shows that b has a minimum when $F = 1$.

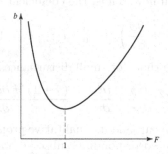

Fig. 6.21. Variation of b with F.

As for the quasi-one-dimensional gas flow, it is interesting to consider flow in a channel whose width first converges and then diverges. From Figure 6.21, we see that it is possible to have a flow that changes from subcritical ($F < 1$) to supercritical ($F > 1$) as long as $F = 1$ at the point of minimum width, and in that case, the depth will be a decreasing function of distance along the channel. As in Section 6.2.3, it may be necessary to introduce a steady hydraulic jump into the supercritical flow in the divergent part of the channel in order to satisfy the downstream conditions.

6.3 *Further Limitations of Linearized Gasdynamics

In this section, we first revisit the linear theory for thin wings in transonic and supersonic two-dimensional steady flow that were derived in Section 4.6.3. We will show how we can (i) discuss the transonic case when $M \simeq 1$ and (ii) extend the supersonic theory to the far field. We then give a brief description of how to deal with shocks that are extreme enough to drive the gas out of thermodynamic equilibrium and hence invalidate the gas law (2.9). Finally, we return to study the flow past a thin wing when $M \gg 1$.

6.3.1 Transonic Flow

In Section 4.6.3, we have already anticipated trouble with our thin wing theories when the free-stream Mach number M is close to unity. To derive a small disturbance theory that is valid in the transonic regime, we have to return to (5.17). Writing the velocity potential as $Ux + \phi$, (5.17) becomes

$$\left(c^2 - \left(U + \frac{\partial \phi}{\partial x} \right)^2 \right) \frac{\partial^2 \phi}{\partial x^2} - 2 \left(U + \frac{\partial \phi}{\partial x} \right) \frac{\partial \phi}{\partial y} \cdot \frac{\partial^2 \phi}{\partial x \partial y} + \left(c^2 - \left(\frac{\partial \phi}{\partial y} \right)^2 \right) \frac{\partial^2 \phi}{\partial y^2} = 0,$$

where

$$c^2 = c_0^2 + \frac{\gamma - 1}{2} U^2 - \frac{(\gamma - 1)}{2} \left(\left(U + \frac{\partial \phi}{\partial x} \right)^2 + \left(\frac{\partial \phi}{\partial y} \right)^2 \right).$$

We can still assume that ϕ and its derivatives are small, but we must remember that $c_0^2 - U^2$ is also small now. Thus, the coefficient of $\partial^2 \phi / \partial x^2$ is

$$c^2 - \left(U + \frac{\partial \phi}{\partial x} \right)^2 \simeq c_0^2 - U^2 - (\gamma + 1) U \frac{\partial \phi}{\partial x},$$

and we are led to deduce that, for small disturbances in transonic flow,

$$(1 - M^2) \frac{\partial^2 \phi}{\partial x^2} + \frac{\partial^2 \phi}{\partial y^2} = \frac{(\gamma + 1)M}{c_0} \frac{\partial \phi}{\partial x} \frac{\partial^2 \phi}{\partial x^2}. \tag{6.62}$$

This equation is nonlinear, and its qualitative properties are much less well understood than are the corresponding subsonic and supersonic equations. The basic difficulty is that when $M = 1$, (6.62) changes from being hyperbolic to elliptic when $\partial \phi / \partial x$ changes sign. This equation is said to be of *mixed type*; some idea of the possible behavior of solutions of such equations can be gained by considering the linear mixed-type Tricomi equation as in Exercise 4.1 (see Garabedian [22]).

We should also mention that we must be prepared for solutions of (6.62) to contain shocks. Indeed, experimental evidence for slightly subsonic flow past a thin wing shows both a supersonic region and a shock, as illustrated in Figure 6.22. Fortunately, these shocks are weak enough for the entropy jump across them to be neglected and hence for the assumption of irrotationality to be justified.

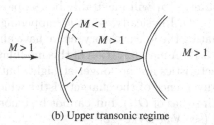

(a) Lower transonic regime

(b) Upper transonic regime

Fig. 6.22. Transonic flow past a thin wing.

6.3.2 The Far Field for Flow past a Thin Wing

The linearized theory of Section 4.6.3 showed that in supersonic flow past a thin wing, the disturbance is confined between characteristics emanating from the leading and trailing edges of the body and that this inevitably entailed some sort of discontinuity across these characteristics. A calculation can be performed to show that weak shocks or expansion fans will, in fact, lie near these characteristics. However, such discontinuities may be neglected when calculating the aerodynamic forces. However, from an environmental point of view, the effect of these discontinuities needs to be understood in the "far field" of the wing in order to assess noise at ground level due to supersonic aircraft.

We first need to assess the region of validity of the linearized solution (4.53), and so we go to the next term in the expansion for the velocity potential by writing $\phi \sim Ux + lU(\varepsilon\phi_0 + \varepsilon^2\phi_1 + \cdots)$ in (5.17). Then, ϕ_0 will be given by (4.53) so that above the wing,

$$\phi_0 = -B^{-1}f_+(x - By) \quad \text{for} \quad y > 0,$$

where $B^2 = U^2/c_0^2 - 1$ and x and y have been made dimensionless with l. The equation for ϕ_1 is

$$B^2\frac{\partial^2\phi_1}{\partial x^2} - \frac{\partial^2\phi_1}{\partial y^2} = -M^2\left(2\frac{\partial\phi_0}{\partial y}\frac{\partial^2\phi_0}{\partial x\partial y} + (\gamma+1)\frac{\partial\phi_0}{\partial x}\frac{\partial^2\phi_0}{\partial x^2} + (\gamma-1)\frac{\partial\phi_0}{\partial x}\frac{\partial^2\phi_0}{\partial y^2}\right).$$

Now, substituting for ϕ_0 and changing to variables $\xi = x - By$ and $\eta = x + By$ leads to

$$4B^2\frac{\partial^2\phi_1}{\partial\xi\partial\eta} = -\frac{M^4}{B^2}(\gamma+1)f_+'(\xi)f_+''(\xi)$$

in $y > 0$, and the solution is

$$\phi_1 = -\frac{M^4}{8B^4}(\gamma + 1)\eta f'_+(\xi)^2 + F(\xi) + G(\eta),$$

where F and G are arbitrary functions. Thus, as x and y increase along the lines where ξ is constant, $\varepsilon^2\phi_1$ will inevitably be comparable in size with $\varepsilon\phi_0$ when x and y are $O(\varepsilon^{-1})$. Physically, what is happening is that nonlinearity is, inexorably, modulating the linear theory, as we have already seen in Section 5.2.4 and in Exercise 5.10. A manifestation of this modulation is that to second order in ε, the characteristics are no longer straight, but curved, as sketched in Figure 6.23. The divergence of the characteristics which intersect the wing is negligible on length scales of $O(l)$, but cannot be ignored in the far field at distances of $O(\varepsilon^{-1}l)$ (see Van Dyke [28]).

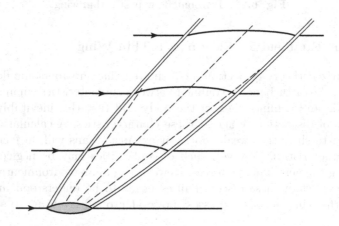

Fig. 6.23. The farfield flow past a thin wing in supersonic flow.

To get a quantitative description of the far-field flow, we can use the same method as that of Section 5.2.4. Guided by the analysis above, we change to variables $\xi = x - By$ and $Y = \varepsilon y$ in (5.17) before writing $\phi = Ux + lU\varepsilon\Phi$ which leads directly to

$$2B\frac{\partial^2\Phi}{\partial\xi\partial Y} = -(\gamma + 1)M^4\frac{\partial\Phi}{\partial\xi}\frac{\partial^2\Phi}{\partial\xi^2} + O(\varepsilon).$$

Thus, the perturbation velocity $u = \partial\Phi/\partial\xi$ satisfies the now familiar kinematic wave equation

$$\frac{\partial u}{\partial Y} + \frac{(\gamma + 1)M^4}{2B}u\frac{\partial u}{\partial\xi} = 0, \tag{6.63}$$

with the initial condition $u = -(1/B)f'(\xi)$ on $Y = 0$. We can even write down the explicit solution (shock waves and all!) in the case where the wing profile is parabolic and at zero incidence (Exercise 6.22). In this case, both the leading and trailing characteristics are weak shocks[6] and this solution reveals the famous "N-wave" solution, which can be shown to be the "generic" solution as $Y \to \infty$ and gives a pressure profile as shown in Figure 6.24 (see Whitham [23]). The shocks weaken as $O(Y^{-1/2})$ at the same time as the expansion wave spreads parabolically in ξ. This explains the "double bang" that is sometimes heard on the ground when an aircraft flies supersonically at altitude, but possibly many kilometers away horizontally.

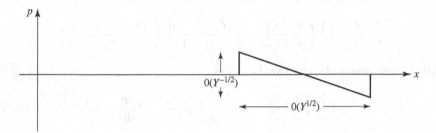

Fig. 6.24. The pressure in the far field.

6.3.3 Non-equilibrium Effects

It is possible for a strong shock to cause a gas such as air to cease to be in thermodynamic equilibrium. There are several mechanisms that can be involved, the commonest being dissociation (in which the molecules are split) and ionization. The latter takes us into the realms of plasma physics, so here we give a brief account of a simple model for dissociation. In both cases, the spirit of the modeling is the same, with the gas being regarded as having two temperatures: its equilibrium temperature T and an *internal temperature T_i*. The temperature T_i measures the energy in the molecular vibrational state of the dissociating gas and is only equal to T in thermodynamic equilibrium. The state furthest from equilibrium is when $T_i = 0$ and this is called the *frozen* state.

As in traditional models for chemical reactions, we postulate a rate equation

$$\tau \frac{dT_i}{dt} = (T - T_i), \tag{6.64}$$

which governs the *relaxation* of the internal temperature to its equilibrium value; τ is the timescale for this relaxation process. We also assume that the internal energy is given by

$$e = c_{v_f} T + c_{v_i} T_i, \tag{6.65}$$

---

[6] Depending on the aerofoil shape and angle of incidence, the leading edge could emit either two weak shock waves or a weak shock and a weak expansion wave.

where c_{v_f} is the specific heat of the frozen gas and $c_{v_f} + c_{v_i} = c_{v_e}$ is the specific heat in equilibrium. When we use (6.64) and (6.65) with the perfect gas law (2.9) to eliminate T and T_i, we see that

$$\left(\tau \frac{d}{dt} + 1\right) e = \frac{\tau}{\gamma_f - 1} \frac{d}{dt}\left(\frac{p}{\rho}\right) + \frac{1}{\gamma_e - 1} \frac{p}{\rho}, \tag{6.66}$$

where γ_e and γ_f are defined by $c_{v_f} = R/(\gamma_f - 1)$ and $c_{v_e} = R/(\gamma_e - 1)$. Now, we can use the energy equation (2.8) with (6.66) to get

$$\left(\tau \frac{d}{dt} + 1\right)\left(\frac{p}{\rho^2} \frac{d\rho}{dt}\right) = \frac{d}{dt}\left(\frac{\tau}{\gamma_f - 1} \frac{d}{dt}\left(\frac{p}{\rho}\right) + \frac{1}{\gamma_e - 1} \frac{p}{\rho}\right).$$

After some manipulation, this becomes

$$\rho\bar{\tau} \frac{d}{dt}\left(\frac{1}{\rho}\left(\frac{dp}{dt} - c_f^2 \frac{d\rho}{dt}\right)\right) + \frac{dp}{dt} - c_e^2 \frac{d\rho}{dt} = 0, \tag{6.67}$$

where $c_f = \sqrt{\gamma_f p/\rho}$ and $c_e = \sqrt{\gamma_e p/\rho}$ are the speeds of sound in the frozen and equilibrium gas respectively, and $\bar{\tau} = ((\gamma_e - 1)/(\gamma_f - 1))\tau$.

As in conventional gasdynamics, it is easiest to see the general properties of the solution for acoustic waves and we also restrict our discussion here to one dimension for simplicity. Following our usual linearizing procedure, we write (2.6) and (2.7) in the form (3.1) and (3.2), respectively, and then use (6.67) to get

$$\bar{\tau} \frac{\partial}{\partial t}\left(\frac{\partial^2 u}{\partial t^2} - c_{f_0}^2 \frac{\partial^2 u}{\partial x^2}\right) + \frac{\partial^2 u}{\partial t^2} - c_{e_0}^2 \frac{\partial^2 u}{\partial x^2} = 0, \tag{6.68}$$

where c_{f_0} and c_{e_0} are the undisturbed speeds of sound. Thus, the model has reduced to a generalized wave equation in which, for small times, the first term dominates and waves propagate with speed c_{f_0}, but eventually these waves transform into ones traveling with speed c_{e_0}. The equilibrium speed c_{e_0} will always be less than the frozen speed c_{f_0} since $c_{v_e} > c_{v_f}$.

Just as in Section 6.3.2, nonlinearity, however small, will ultimately modulate many linear propagating waves. A somewhat tedious generalization of the argument leading to (6.63) shows that when relaxation occurs, the very far field of a wave is described by Burgers' equation (6.11), (as in Exercise 5.10).

6.3.4 Hypersonic Flow

In Section 4.6.3, we drew attention to the fact that the linearized approximation is no longer valid when $M\varepsilon = O(1)$. When this happens, the slope of the characteristic at the leading edge of the wing is comparable with the slope of the body; hence, the shock that is generated there will no longer be weak.

Thus, we need to reconsider the shock relations in this situation in order to determine the magnitudes of the various flow quantities between the shock and the body.

Suppose a two-dimensional steady shock is generated by flow past a symmetric wedge as illustrated in Figure 6.25, where 2ε is the small angle of the wedge and α, the angle of inclination of the shock, is also small. Now, from the Rankine–Hugoniot condition (6.46), with $\beta = \alpha$ and $\theta = \varepsilon$, the assumption that α and ε are both small implies that

$$(\gamma + 1)M^2\alpha(\alpha - \varepsilon) = 2 + (\gamma - 1)M^2\varepsilon^2,$$

and, hence,

$$\alpha = \frac{\gamma + 1}{4}\varepsilon\left(1 + \left[1 + \left(\frac{4}{(\gamma + 1)\varepsilon M}\right)^2\right]^{1/2}\right).$$

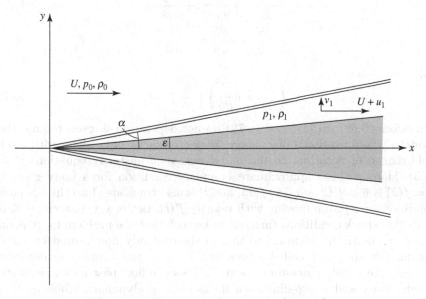

Fig. 6.25. Hypersonic flow past a thin wedge.

Thus, α is of $O(\varepsilon)$ as long as $M\varepsilon \geq O(1)$. On the body, the boundary condition is

$$v_1 = (U + u_1)\varepsilon,$$

and parallel to the shock, the condition (6.36) gives

$$u_1 \cos\alpha + v_1 \sin\alpha = 0,$$

so that we can deduce that $v_1/U = O(\varepsilon)$ and $u_1/U = O(\varepsilon^2)$. Furthermore, from the Rankine–Hugoniot conditions (6.43) and (6.44), it is easy to show that

$$p_1 - p_0 = O(\rho_0 U^2 \varepsilon^2) \quad \text{and} \quad \frac{\rho_1}{\rho_0} = O(1).$$

Using these estimates, we can now scale the variables in the flow between the shock and a more general body whose slope is of $O(\varepsilon)$ by writing

$$u = U(1 + \varepsilon^2 \bar{u}), \qquad v = U\varepsilon\bar{v}, \qquad p = p_0 + \rho_0 U^2 \varepsilon^2 \bar{p},$$
$$\rho = \rho_0 \bar{\rho}, \qquad y = \varepsilon l \bar{y}, \qquad x = l\bar{x},$$

where l is the streamwise lengthscale of the body. Then, we can expect that the barred variables will all be $O(1)$ behind the shock and the first approximation to (2.6)–(2.8) will be

$$\frac{\partial \bar{\rho}}{\partial \bar{x}} + \frac{\partial}{\partial \bar{y}}(\bar{\rho}\bar{v}) = 0, \tag{6.69}$$

$$\frac{\partial \bar{u}}{\partial \bar{x}} + \bar{v}\frac{\partial \bar{u}}{\partial \bar{y}} = -\frac{1}{\bar{\rho}}\frac{\partial \bar{p}}{\partial \bar{x}}, \tag{6.70}$$

$$\frac{\partial \bar{v}}{\partial \bar{x}} + \bar{v}\frac{\partial \bar{v}}{\partial \bar{y}} = -\frac{1}{\bar{\rho}}\frac{\partial \bar{p}}{\partial \bar{y}} \tag{6.71}$$

and

$$\left(\frac{\partial}{\partial \bar{x}} + \bar{v}\frac{\partial}{\partial \bar{y}}\right)\left(\frac{\bar{p}}{\bar{\rho}^\gamma}\right) = 0. \tag{6.72}$$

Equations (6.69), (6.71), and (6.72) do not depend on $\bar{u}$, even though they model a two-dimensional flow; hence, they are exactly equivalent, with a suitable change of notation, to the *one-dimensional* unsteady equations (5.3)–(5.5). Moreover, the approximate boundary condition for a body given by $\bar{y} = f(\bar{x})$ is $\bar{v} = f'(\bar{x})$ on $\bar{y} = f(\bar{x})$, and this also translates into the boundary condition for a piston moving with velocity $f'(t)$. Better yet, Exercise 6.23 reveals that shock conditions turn out to be such that the problem for $\bar{p}$, $\bar{\rho}$, and $\bar{v}$ is mathematically *identical* to that of the unsteady flow caused by such a piston. This analogy is called *hypersonic similitude* and a similar analog exists between the steady three-dimensional hypersonic flow past an axisymmetric slender body and a two-dimensional unsteady gasdynamics problem. Thus, we have found a class of gasdynamic flows in which time and space can be identified.

If $M\varepsilon \gg 1$, then the shock is strong enough for the shock relations to simplify still further. Suppose, in particular, that we consider such a flow past a two-dimensional power-law body given by $\bar{y} = C\bar{x}^k$. Thus, as shown in Exercise 6.24, the shock relations are simple enought for us to find a similarity solution which depends only on the variable $\bar{y}/\bar{x}^k$. The principle of hypersonic similitude now states that the problem of a strong shock driven by a piston whose position is proportional to t^k also reduces to ordinary differential equations in the single variable x/t^k; happily, a similar reduction to ordinary differential equations also occurs for radially or spherically symmetric pistons whose position is proportional to t^k.

A dramatic example of such an unsteady flow which can be solved explicitly is that of the *blast wave* caused by a sudden explosion at $t = 0$ at a point O. Such an explosion results in the sudden release of a large amount of energy E and causes a strong shock to expand into the initially quiescent surrounding gas. If the shock position is given by $r = R(t)$ and the density, pressure, and velocity of the gas just behind the shock are ρ_1, p_1, and v_1 respectively, then since the shock is strong, the Rankine–Hugoniot conditions (6.24)–(6.26) reduce to

$$\rho_1 = \rho_0 \frac{(\gamma + 1)}{(\gamma - 1)},$$

$$v_1 = \frac{2\dot{R}}{\gamma + 1}$$

and

$$p_1 = \frac{2\rho_0 \dot{R}^2}{\gamma + 1},$$

where ρ_0 is the ambient density. Thus, the only physical quantities that enter this problem are r, t, E, and ρ_0. In the three-dimensional case, the only nondimensional quantity that can be constructed from these quantities is

$$\xi = r \left(\frac{\rho_0}{Et^2} \right)^{1/5}.$$

Hence, ρ/ρ_0 will be a function of ξ, and since ρ/ρ_0 is constant at the shock, we retrieve the famous result that an atomic bomb blast grows such that $R/t^{2/5} = K = $ constant. In order to find K, we must solve the equations of unsteady gas flow behind the shock, which we know reduce to ordinary differential equations in ξ with boundary conditions given on the shock. One integral of the motion can be found immediately by observing that the total energy behind the shock remains constant and equal to E, and this leads us to a relation between E and K. Thus, one can estimate the energy release in an atomic explosion by observing the growth of the emitted blast wave. This theory also reveals that such blast waves could be generated by a spherical piston expanding so that the radius grows like $t^{2/5}$.

Going down a dimension, we can consider a cylindrical blast wave caused by a line explosion releasing a fixed amount of energy per unit length. Then, the dimensional argument shows that the blast wave expands so that its radius is proportional to $t^{1/2}$. Using hypersonic similitude, we can deduce that the same solution holds for the hypersonic flow past a slender axisymmetric body whose radius is proportional to $x^{1/2}$. There is one caveat: this body has a blunt nose and the flow near the nose will not be susceptible to hypersonic small disturbance theory. An asymptotic analysis reveals that there is a thin *entropy layer* near the body where a new solution is needed, but, luckily, this does not affect the position of the shock to first order. For this case, it is also possible to show that the energy release in the cylindrical blast wave is directly related to the drag on the nose of the blunt body in hypersonic flow (Anderson [29]).

Happily, we can end our discussion of nonlinear waves on a less warlike note by remarking that high-speed ocean transportation can benefit from a very similar kind of similitude. As shown in Exercise 6.25, the waves caused by a high-speed (high Froude number) slender ship moving with constant velocity are equivalent to the unsteady waves produced by a wavemaker at the end of a one-dimensional water tank. Whereas hypersonic similitude demands that $M\varepsilon = O(1)$, it transpires that the slender ship theory is valid when $F\varepsilon^2 = O(1)$, where F is the Froude number and ε is the slenderness parameter.

Exercises

R6.1 Derive (6.21) from (6.16) and deduce that for any function f,

$$[f\rho u] - \dot{X}[f\rho] = [f]m,$$

where $m = \rho(u - \dot{X})$. Then, put $f = \dot{X}$ to derive (6.22) from (6.18). Finally, put $f = p/\rho$ to show that (6.20) implies that

$$\frac{m\gamma}{\gamma - 1}\left[\frac{p}{\rho}\right] + \left[\frac{1}{2}\rho u^3\right] + \dot{X}[p] = \dot{X}\left[\frac{1}{2}\rho u^2\right].$$

Using (6.22) for $[p]$ and writing $\left[\frac{1}{2}\rho u^2(u - \dot{X})\right] = m\left[\frac{1}{2}u^2\right]$, deduce (6.23).

R6.2 Writing $M_i = (\dot{X} - u_i)/c_i$, show that (6.21) implies that $[\rho^2 M^2 c^2] = 0$, where $c^2 = p/\rho$. Deduce that $p_2\rho_2/p_1\rho_1 = M_1^2/M_2^2$.

Show that (6.22) implies that $[p(1 + \gamma M^2)] = 0$ and, hence, that

$$\frac{p_2}{p_1} = \frac{1 + \gamma M_1^2}{1 + \gamma M_2^2}.$$

Show that (6.23) implies that $[(p/\rho)(1 + [(\gamma - 1)/2]M^2)] = 0$ and, hence, that

$$\frac{p_2\rho_1}{p_1\rho_2} = \frac{1 + [(\gamma - 1)/2]M_1^2}{1 + [(\gamma - 1)/2]M_2^2}.$$

Combine these results to show that

$$(M_2^2 - M_1^2)[M_2^2(2\gamma M_1^2 - (\gamma - 1)) - ((\gamma - 1)M_1^2 + 2)] = 0$$

and, hence, derive (6.24)–(6.26).

R6.3 Show that the entropy jump across a shock is

$$[S] = c_V \left[\log \frac{p}{\rho^\gamma} \right].$$

Use (6.24)–(6.26) to write this as

$$[S] = c_V \left(\log(2\gamma x - (\gamma - 1)) + \gamma \log \left(\frac{2}{x} + \gamma - 1 \right) - (\gamma + 1) \log(\gamma + 1) \right),$$

where $x = M_1^2$. Show further that

$$\frac{1}{c_V} \frac{d[S]}{dx} = 2\gamma(\gamma - 1) \frac{(x - 1)^2}{(2\gamma x - (\gamma - 1))(2/x + \gamma - 1)x^2}.$$

Deduce the following:
(i) $[S] > 0$ when $x > 1$.
(ii) $[S] = O((x - 1)^3)$ when $x \downarrow 1$.

6.4 From (6.21) and (6.22) show that across a normal shock wave

$$(u_1 - u_2)^2 = (p_2 - p_1) \left(\frac{1}{\rho_1} - \frac{1}{\rho_2} \right),$$

where, as usual, the suffix 1 is ahead of the shock and the suffix 2 is behind.

From (6.24) and (6.25) show that

$$\frac{\rho_2}{\rho_1} = \frac{(\gamma - 1)p_1 + (\gamma + 1)p_2}{(\gamma + 1)p_1 + (\gamma - 1)p_2}.$$

Now, suppose that a normal shock with pressure ratio $\pi(= p_2/p_1)$ is reflected from a plane parallel rigid boundary. Show that the pressure ratio of the reflected shock is p, where

$$p = \frac{(3\gamma - 1)\pi - (\gamma - 1)}{(\gamma - 1)\pi + (\gamma + 1)}.$$

6.5 A plane shock wave is reflected from a parallel plane wall. Show that if U_1 is the speed of approach and U_2 the speed of departure, then

$$(U_1 - u)^2 - c^2 = -\left(\frac{\gamma + 1}{2} \right) u(U_1 - u),$$

$$(U_2 + u)^2 - c^2 = \frac{\gamma + 1}{2} u(U_2 + u),$$

where u and c are the velocity and speed of sound, respectively, behind the incoming shock. Deduce that

$$U_1 > U_2$$

as long as $\gamma < 3$.

R6.6 Rearrange (6.46) to show that

$$\tan \theta = 2 \cot \beta \frac{M_1^2 \sin^2 \beta - 1}{(\gamma + \cos 2\beta)M_1^2 + 2}.$$

Show that $\theta = 0$ when $\beta = \pi/2$ and when $\beta = \sin^{-1}(1/M_1)$. Deduce that for fixed M_1, $\theta(\beta)$ has at least one maximum (which can be shown to be close to, but not quite equal to, the value of β for which $M_2 = 1$). Show also that when θ is small and β is near $\sin^{-1}(1/M_1)$,

$$\theta \sim \frac{4(M_1^2 - 1)}{(\gamma + 1)M_1^2} \left(\beta - \sin^{-1} \frac{1}{M_1} \right).$$

Deduce that, for such weak shocks, the downstream pressure, Mach number, and velocity are given by

$$\frac{p_2}{p_1} = 1 + \frac{\gamma M_1^2 \theta}{\sqrt{M_1^2 - 1}},$$

$$\frac{M_2}{M_1} = 1 - \theta \left(1 + \frac{\gamma - 1}{2} M_1^2 \right) \left(\sqrt{M_1^2 - 1} \right)^{-1}$$

and

$$\frac{u_2}{u_1} = 1 - \frac{\theta}{\sqrt{M_1^2 - 1}},$$

respectively, where u_1 is the velocity in front of the shock and u_2 is the component of the velocity in the same direction behind the shock.

*6.7 A gas with speed u_1 and sound speed c_1 flows homentropically to a state where the Mach number is unity and its sound speed is c_*, which is called the *critical speed of sound*. Show that

$$(\gamma + 1)c_*^2 = (\gamma - 1)u_1^2 + 2c_1^2$$

and hence use (6.42) to show that the critical speeds of sound on either side of an oblique shock are the same.

Suppose a gas with velocity $(u_1, 0)$ and sound speed c_1 passes through a shock, inclined at an angle β to the x axis, to a state with velocity (u_2, v_2): Show that

$$u_1 \cos \beta = u_2 \cos \beta + v_2 \sin \beta,$$
$$\rho_1 u_1 \sin \beta = \rho_2(u_2 \sin \beta - v_2 \cos \beta),$$
$$p_1 + \rho_1 u_1^2 \sin^2 \beta = p_2 + \rho_2(u_2 \sin \beta - v_2 \cos \beta)^2$$

and

$$\frac{1}{2} u_1^2 \sin^2 \beta + \frac{c_1^2}{\gamma - 1} = \frac{1}{2}(u_2 \sin \beta - v_2 \cos \beta)^2 + \frac{c_2^2}{\gamma - 1}$$
$$= \frac{\gamma + 1}{2(\gamma - 1)} c_*^2 - \frac{1}{2} u_1^2 \cos^2 \beta.$$

Eliminate p_1, ρ_1, p_2, ρ_2, and β from these equations to show that

$$v_2^2 = \frac{(u_1 - u_2)^2(u_1 u_2 - c_*^2)}{([2/(\gamma+1)]u_1^2 + c_*^2 - u_1 u_2)}.$$

Sketch this curve in the (u_2, v_2) plane for given u_1 and c_* and show that only the segment $c_*^2/u_1 < u_2 < u_1$ is physically relevant. Confirm that there are two values of (u_2, v_2) for any given value of v_2/u_2 less than the maximum possible deflection.

The curve in the (u_2, v_2) plane is called the *shock polar*.

6.8 Show that if the Prandtl–Meyer expansion (5.65)–(5.66) is weak, the flow deflection θ, which is now negative, is approximately related to the downstream Mach number M_2 by

$$\theta \sim (\mu_1 - \mu) f'(\mu_1)$$
$$\sim \frac{(M_2 - M_1)}{M_1 \sqrt{M_1^2 - 1}} \left(1 - \frac{(\gamma+1)M_1^2}{2 + (\gamma-1)M_1^2}\right).$$

Deduce that

$$\frac{M_2}{M_1} \sim 1 - \theta \left(1 + \frac{\gamma-1}{2}M_1^2\right)\left(\sqrt{M_1^2 - 1}\right)^{-1},$$

which is the same as the formula obtained for a weak shock in Exercise 6.6.

However, remember θ is now negative and M_2 is now greater than M_1 and that weak shocks and weak expansions refract the flow in *opposite* directions.

6.9 Suppose that a weak shock making an angle β to the positive x axis impinges on a wall $y = 0$ from below. Using Exercise 6.6, show that a weak shock reflects from $y = 0$, making an angle $-\beta$ with the positive x axis.

This specular reflection is only true for weak shocks. The reflection angle may be greater or less than β depending on the strength of the shock.

6.10 A circular cone is placed in a uniform supersonic stream with its axis parallel to the stream. Show that if the resulting shock wave is a concentric circular cone, then there is a velocity potential Φ such that the radial and transverse velocities between the shock and the body are

$$u_r = \frac{\partial \Phi}{\partial r}, \qquad u_\theta = \frac{1}{r}\frac{\partial \Phi}{\partial \theta},$$

where r and θ are spherical polar coordinates. Given that

$$\nabla \cdot \rho(u_r, u_\theta, 0) = \frac{1}{r \sin\theta}\frac{\partial}{\partial \theta}(\sin\theta \rho u_\theta) + \frac{1}{r^2}\frac{\partial}{\partial r}(r^2 \rho u_r),$$

show that there is a similarity solution in which $\Phi = r\phi(\theta)$ with

$$\left(1 - \frac{1}{c^2}\left(\frac{d\phi}{d\theta}\right)^2\right)\frac{d^2\phi}{d\theta^2} - \left(\frac{\phi}{c^2}\frac{d\phi}{d\theta} - \cot\theta\right)\frac{d\phi}{d\theta} + 2\phi = 0,$$

where
$$c^2 = -\frac{(\gamma - 1)}{2}\left(\phi^2 + \left(\frac{d\phi}{d\theta}\right)^2\right) + \text{constant}.$$

R6.11 Write $F_i = (u_i - \dot{X})/\sqrt{g\eta_i}$ in (6.49)–(6.50) and show that
$$[s^3 F] = [\tfrac{1}{2}s^4 + F^2 s^4] = 0.$$

Deduce that
$$F_1^2 = \frac{s_2^2}{s_1^2}\frac{1 + (s_2^2/s_1^2)}{2}$$

and use (6.52) to infer that $|F_1| \geq 1$ and $|F_2| \leq 1$.

Note the analogy between these inequalities for the Froude number in shallow water theory and the inequalities (6.28) for the Mach number in gasdynamics.

6.12 A bore invades water originally at rest in a straight horizontal channel of uniform rectangular cross section. The depth of the water increases from H to $2H$ by the passage of the bore. Show that the velocity behind the bore is $\sqrt{3gH}$. The bore is reflected at the closed end of the channel. Show that, after reflection, the depth of water at the closed end is $\frac{1}{2}(1+\sqrt{33})H$.

*6.13 Show that for the two-dimensional shallow water equations of Exercise 5.5, the steady shock (bore) relations for the conservation of mass and momentum are
$$\frac{dy}{dx} = \frac{[\eta v]}{[\eta u]} = \frac{[\eta u v]}{[\tfrac{1}{2}g\eta^2 + \eta u^2]} = \frac{[\tfrac{1}{2}g\eta^2 + \eta v^2]}{[\eta u v]}.$$

Show that the energy dissipation across the bore depends on dy/dx and use the result of Exercise 5.5 to deduce that if a uniform stream encounters a curved bore, the downstream vorticity $\partial u/\partial y - \partial v/\partial x$ will be non-zero.

R6.14 Water of depth s_l^2/g is contained in $-\infty < x < 0$ and is separated by a sluice gate from water of depth s_r^2/g in $0 < x < \infty$, where $s_r < s_l$. At time $t = 0$, the sluice gate is suddenly removed. Show that the solution comprises the following:

(i) An expansion fan in $-s_l t < x < (u_1 - s_1)t$.

(ii) A region of uniform flow where $s = s_1$ and $u = u_1$ for $(u_1 - s_1)t < x < Vt$.

(iii) A hydraulic jump at $x = Vt$.

Write down sufficient equations to determine u_1, s_1, and V and show that if $u_1 > 2s_l/3$, and $t > 0$, then the water depth at $x = 0$ is $4s_l^2/9g$ and the discharge rate is $8s_l^3/27g$.

R6.15 Gas flows steadily out of a reservoir, where the density is ρ_0 and the sound speed c_0, into a duct of slowly varying cross-section $A(x)$. The duct area initially decreases to a minimum at $x = X$ and then increases. Show that if the Mach number is M and the mass flow in the duct is Q, then
$$\frac{\rho_0 c_0}{Q}\frac{dA}{dx} = \left(1 - \frac{1}{M^2}\right)\left(1 + \left(\frac{\gamma - 1}{2}\right)M^2\right)^{(3-\gamma)/2(\gamma-1)}\frac{dM}{dx}.$$

Deduce that if the duct is choked so that $M = 1$ at $x = X$, then

$$Q = \left(\frac{2}{\gamma + 1}\right)^{(\gamma+1)/2(\gamma-1)} \rho_0 c_0 A(X).$$

6.16 Water flows into an open channel from a reservoir where the total head is gH. Show that if the channel breadth $b(x)$ decreases to a minimum b_* downstream before increasing again, then if the Froude number F attains the value unity, it does so when $b = b_*$. Prove that in such a choked flow, the flow rate is

$$q = (\tfrac{2}{3})^{3/2} g^{1/2} H^{3/2} b_*.$$

6.17 (i) Show that the density and pressure ratios across the expansion fan generated by a piston moved impulsively out of a tube with velocity U_p, as in Exercise 5.15, are given by

$$\frac{\rho_2}{\rho_1} = \left(1 - \frac{\gamma - 1}{2}\frac{|U_p|}{c_1}\right)^{2/\gamma - 1}, \qquad \frac{p_2}{p_1} = \left(\frac{\rho_2}{\rho_1}\right)^{\gamma},$$

where $|U_p|$ is the piston speed, assumed less than $2c_1/(\gamma - 1)$, and subscripts 1 and 2 refer to conditions ahead of and behind the fan, respectively.

*(ii) Inviscid gas is contained in an infinite shock tube lying along the x axis. An impermeable membrane at $x = 0$ separates gas with pressure p_l and sound speed c_l in $x < 0$ from the same gas at conditions p_r and c_r in $x > 0$, where $p_l > p_r$. At time $t = 0$, the membrane is ruptured. Show that the subsequent flow comprises the following:
   (a) An expansion fan in $-c_l t < x < (V - c_2)t$.
   (b) A uniform flow region in $(V - c_2)t < x < Vt$, in which $u = V$, $p = p_1$, $c = c_2$.
   (c) A uniform flow region in $Vt < x < Ut$, in which $u = V$, $p = p_1$, and $c = c_1$.
   (d) A shock at $x = Ut$, where the unknowns satisfy

$$c_2 = c_l - \frac{\gamma - 1}{2}V, \qquad \frac{p_1}{p_l} = \left(1 - \frac{(\gamma - 1)}{2}\frac{V}{c_l}\right)^{2\gamma/(\gamma-1)},$$

$$(U - V)\rho_1 = U\rho_r,$$

$$p_1 + (U - V)\rho_1^2 = p_r + \rho_r U^2$$

and

$$\frac{c_1^2}{\gamma - 1} + \frac{1}{2}(V - U)^2 = \frac{c_r^2}{\gamma - 1} + \frac{1}{2}U^2,$$

where ρ_r and ρ_1 are the densities ahead of and behind the shock respectively. Show that

$$\frac{V}{c_l} = \frac{2}{\gamma - 1}\left(1 - \left(\frac{p_1}{p_l}\right)^{(\gamma-1)/2\gamma}\right)$$

and that

$$\frac{V}{c_r} = \left(\frac{p_1}{p_r} - 1\right)\left(\frac{2/\gamma}{(\gamma+1)p_1/p_r + (\gamma-1)}\right)^{1/2}$$

and, hence, deduce the *shock tube equation*

$$\frac{p_l}{p_r} = \frac{p_1}{p_r}\left(1 - \frac{(\gamma-1)(c_r/c_l)(p_1/p_r - 1)}{[2\gamma(\gamma+1)p_1/p_r + \gamma - 1)]^{1/2}}\right)^{-2\gamma/(\gamma-1)}.$$

Which of the flow variables is continuous at the contact discontinuity $x = Vt$?

6.18 Suppose that two unequal weak shocks make angles β and β' with a stream of Mach number M_1 as in Figure 6.12.

Show that the deflections satisfy

$$\theta - \phi = -\theta' + \phi'$$

and use the results of Exercise 6.6 to show that

$$\theta + \phi = \theta' + \phi'.$$

Show that the flow deflection is

$$\theta - \theta' = \frac{4(M_1^2 - 1)}{(\gamma+1)M_1^2}(\beta - \beta')$$

and that, in general, there will be a contact discontinuity (in this case, a vortex sheet) separating the downstream flow into parallel gas streams with unequal speeds.

6.19 Suppose that instead of the configurations in Figure 6.12, two weak shocks intersect as in Figure 6.13. Show that the shocks can merge to form a third shock with $\phi = \theta + \theta'$ and that there will again be a contact discontinuity in the downstream flow.

In this situation, it can be shown that when the shocks are stronger, an expansion fan will also be formed near the negative characteristic through the intersection point.

6.20 A weak shock S_i impinges on a vortex sheet ABC which separates two supersonic streams with Mach numbers M_1 and M_1', as shown in Figure 6.14.

Show that if the deflections θ, ϕ, and θ' are measured as shown in Figure 6.14, then using the results of Exercise 6.6,

$$\theta - \phi = \theta'$$

and

$$(\theta + \phi)\frac{M_1^2}{\sqrt{M_1^2 - 1}} = \theta'\frac{M_1'^2}{\sqrt{M_1'^2 - 1}}.$$

Show further that there will be a contact discontinuity in the down-stream flow.

If $M_1 > M_1' > \sqrt{2}$, show that $\phi < 0$, so that the second shock S_2 above the vortex sheet will be replaced by an expansion fan for which the above results still apply (see Exercise 6.8). Show that the strength of the shock T transmitted by the vortex sheet is always less than the strength of the incident shock I.

This idea has been proposed for attenuating sonic boom from supersonic aircraft.

6.21 Suppose gas flows steadily down a slowly-varying channel with walls given by

$$y = \pm S(\varepsilon x),$$

where $\varepsilon \ll 1$. Assuming that $u = O(U)$, $v = O(\varepsilon U)$, $x = O(L)$ and variations in y are of $O(\varepsilon L)$, show that the equations of continuity and momentum are approximated by

$$\frac{\partial}{\partial x}(\rho u) + \frac{\partial}{\partial y}(\rho v) = 0,$$

$$\frac{\partial}{\partial x}(\rho u^2) + \frac{\partial}{\partial y}(\rho u v) + \frac{\partial p}{\partial x} = 0,$$

$$\frac{\partial p}{\partial y} = 0,$$

with

$$V = \pm \varepsilon u S' \quad \text{on } y = \pm S(\varepsilon x).$$

Show that

$$\frac{d}{dx}\int_{-S}^{S} \rho u \, dy = 0$$

and

$$\frac{d}{dx}\int_{-S}^{S} \rho u^2 \, dy + 2S\frac{dp}{dx} = 0.$$

Assuming additionally that the flow is, to lowest order, irrotational and homentropic, show that u, ρ, and p are all approximately functions of x alone and that their averages over the channel width satisfy

$$\bar{\rho}\bar{u}S = \text{constant}, \qquad \frac{1}{2}\bar{u}^2 + \frac{\gamma\bar{p}}{(\gamma-1)\bar{\rho}} = \text{constant}, \qquad \bar{p}/\bar{\rho}^\gamma = \text{constant.},$$

where $\bar{\rho} = (1/2S)\int_{-S}^{S}\rho\,dy$ and similarly for $\bar{p}$ and $\bar{u}$.

*6.22 (i) Show that the far field of a supersonic stream past a thin wing is modeled, in $Y > 0$, by

$$\frac{\partial u}{\partial Y} + \frac{\gamma+1}{2B}M^4 u\frac{\partial u}{\partial \xi} = 0$$

in the notation of Section 6.3.2, with $u = -(1/B)f'(\xi)$ on $Y = 0$.

Show also that the Rankine–Hugoniot condition for this equation is that the leading-edge shock slope in (ξ, Y) coordinates is

$$\frac{d\xi}{dY} = \frac{(\gamma+1)M^4}{4B}\frac{[u^2]}{[u]} = \frac{(\gamma+1)M^4}{4B}u_+,$$

where u_+ is the value of u just downstream of the leading shock. Check this result by using Exercise 6.6 to show that

$$\varepsilon u_+ = -\frac{4B}{(\gamma+1)M^2}(\beta - \sin^{-1}\frac{1}{M}),$$

where $d\xi/dY = (1/\varepsilon)(\cot\beta - B)$.

(ii) Suppose the wing is such that $f(\xi) = l^2 - \xi^2$ for $-l < \xi < l$. Show that the leading-edge shock wave is given by

$$\frac{d\xi}{dY} = \frac{(\gamma+1)M^4}{4B}u_0(\xi_0),$$

where $u_0(\xi_0)$ is the value of u on the characteristic

$$\frac{d\xi}{dY} = \frac{(\gamma+1)M^4}{2B}u_0(\xi_0),$$

with $\xi = \xi_0$ when $Y = 0$. Deduce that

$$\xi_0 = \frac{\xi}{(1 + ((\gamma+1)M^4/B)Y)}$$

and, hence, show that the shock wave is the parabola

$$\frac{(\gamma+1)M^4Y}{B} + 1 = \alpha\xi^2$$

for some constant α. Consider the flow as $Y \downarrow 0$ to show that $\alpha = 1/l$.
Remark: A similar calculation reveals the existence of a parabolic shock from the trailing edge, so that the far field pressure is an "N-wave."

*6.23 (i) Show that if β and θ are small, (6.44) implies that

$$\theta \sim \frac{2\beta}{\gamma+1}\left(1 - \frac{1}{M_1^2\beta^2}\right). \tag{6.73}$$

(ii) In hypersonic flow, the variables $(\bar{p}, \bar{\rho}, \bar{v}, \bar{x}, \bar{y})$ in (6.69), (6.71), and (6.72) are identified with the variables (p, ρ, u, t, x) in the one-dimensional unsteady isentropic gasdynamic equations.
   If a piston is pushed into a gas at rest with pressure p_1 and density ρ_1, show from (6.24) and (6.25) that the pressure p_2, velocity u_2, and

density ρ_2 just behind the shock, whose position is given by $x = x_s(t)$, are given by

$$\frac{p_2}{p_1} = \frac{2\gamma M_1^2 - (\gamma - 1)}{(\gamma + 1)}, \qquad \frac{\rho_2}{\rho_1} = \frac{(\gamma + 1)M_1^2}{2 + (\gamma - 1)M_1^2},$$

$$u_2 = \frac{2\dot{x}_s}{\gamma + 1}\left(1 - \frac{1}{M_1^2}\right),$$

where $M_1 = \dot{x}_s/c_1$ and $c_1^2 = \gamma p_1/\rho_1$.

For the hypersonic problem, show that if the shock is given by $\bar{y} = Y_s(\bar{x})$, then $\beta = \varepsilon Y_s'$, and from (6.43) and (6.44), the shock relations are

$$\frac{p_2}{p_1} = \frac{2\gamma M_1^2 \beta^2 - (\gamma - 1)}{\gamma + 1} \quad \text{and} \quad \frac{\rho_2}{\rho_1} = \frac{(\gamma + 1)M_1^2 \beta^2}{2 + (\gamma - 1)M_1^2 \beta^2};$$

deduce from (6.73) above that, at the shock,

$$\bar{v} = \frac{2Y_s'}{\gamma + 1}\left(1 - \frac{1}{M_1^2 \beta^2}\right).$$

Hence, complete the identification that leads to the principle of hypersonic similitude.

*6.24 Show that if gas streams with Mach number M past a thin wing with slope $O(\varepsilon)$ and $M\varepsilon \gg 1$, then, in the notation of (6.69)–(6.72) and using the results of Exercise 6.23, the shock conditions on $\bar{y} = \bar{Y}_s(\bar{x})$ are

$$\bar{p} = \frac{2}{\gamma + 1}\bar{Y}_s'^{1/2}, \qquad \bar{\rho} = \frac{\gamma + 1}{\gamma - 1} \quad \text{and} \quad \bar{v} = \frac{2\bar{Y}_s'}{\gamma + 1}.$$

Deduce that if the wing is $\bar{y} = b\bar{x}^k$ for $\bar{x} > 0$, then there is a similarity solution

$$\bar{Y}_s = s\bar{x}^k, \qquad \bar{p} = \bar{x}^{2(k-1)}P(\zeta), \qquad \bar{\rho} = R(\zeta), \qquad \bar{v} = \bar{x}^{k-1}V(\zeta),$$

where $\zeta = \bar{y}/\bar{x}^k$, which satisfies

$$-k\gamma R' + (RV)' = 0,$$
$$(k-1)V - k\gamma V' + VV' = -P'/R$$

and

$$\left(2k - 2 + (V - k\zeta)\frac{d}{d\zeta}\right)(P/R^\gamma) = 0$$

with

$$P(s) = \frac{2k^2 s^2}{\gamma + 1}, \qquad R(s) = \frac{\gamma + 1}{\gamma - 1}, \qquad V(s) = \frac{2ks}{\gamma + 1},$$

and

$$V(b) = kb.$$

*6.25 The equation of a ship moving in the z direction with velocity V is given by $F(x, y, z - Vt) = 0$. Show that in steady flow, with $\xi = z - Vt$, the potential for the flow generated by the passage of the ship satisfies

$$\frac{\partial^2 \phi}{\partial x^2} + \frac{\partial^2 \phi}{\partial y^2} + \frac{\partial^2 \phi}{\partial \xi^2} = 0,$$

with

$$-V\frac{\partial \phi}{\partial \xi} + g\eta + \frac{1}{2}\left(\left(\frac{\partial \phi}{\partial x}\right)^2 + \left(\frac{\partial \phi}{\partial y}\right)^2 + \left(\frac{\partial \phi}{\partial \xi}\right)^2\right) = 0$$

and

$$\frac{\partial \phi}{\partial y} = -V\frac{\partial \eta}{\partial \xi} + \frac{\partial \phi}{\partial x}\frac{\partial \eta}{\partial x} + \frac{\partial \phi}{\partial \xi}\frac{\partial \eta}{\partial \xi}$$

on the free surface $y = \eta$ and

$$V\frac{\partial F}{\partial \xi} = \frac{\partial \phi}{\partial x}\frac{\partial F}{\partial x} + \frac{\partial \phi}{\partial y}\frac{\partial F}{\partial y} + \frac{\partial \phi}{\partial \xi}\frac{\partial F}{\partial \xi}$$

on the ship $F(x, y, \xi) = 0$; also, $|\nabla\phi| \to 0$ at infinity since there are no incoming waves.

Now suppose the ship is narrow and of length l, so that $F(x, y, \xi) = \tilde{F}(X, Y, \xi)$, where $x = \varepsilon l X$, $y = \varepsilon l Y$, and $\xi = l\zeta$. Also, suppose that the Froude number is so large that $\varepsilon V^2/gl = f = O(1)$ as $\varepsilon \to 0$. Show that if $\phi = \varepsilon^2 l V \tilde{\phi}$, $\eta = \varepsilon l \tilde{\eta}$, then, to lowest order in ε,

$$\frac{\partial^2 \tilde{\phi}}{\partial X^2} + \frac{\partial^2 \tilde{\phi}}{\partial Y^2} = 0$$

with, on $Y = \tilde{\eta}$,

$$-\frac{\partial \tilde{\phi}}{\partial \zeta} + f^{-1}\tilde{\eta} + \frac{1}{2}\left(\left(\frac{\partial \tilde{\phi}}{\partial X}\right)^2 + \left(\frac{\partial \tilde{\phi}}{\partial Y}\right)^2\right) = 0$$

and

$$\frac{\partial \tilde{\phi}}{\partial Y} = \frac{\partial \tilde{\eta}}{\partial \zeta} + \frac{\partial \tilde{\phi}}{\partial X}\frac{\partial \tilde{\eta}}{\partial X}$$

and, on $\tilde{F} = 0$,

$$\frac{\partial \tilde{F}}{\partial \zeta} = \frac{\partial \tilde{\phi}}{\partial X}\frac{\partial \tilde{F}}{\partial X} + \frac{\partial \tilde{\phi}}{\partial Y}\frac{\partial \tilde{F}}{\partial Y}.$$

Show that when $-\zeta$ is identified with time t, these are the equations of surface gravity waves in two dimensions, driven by a surface penetrating wavemaker $\tilde{F}(X, Y, -t) = 0$.

7

Epilogue

As explained in Chapter 1, this book has changed the emphasis of its progenitor "Inviscid Fluid Flows" [1] by reorganizing much of the material in such a way that the applicability of the analysis can be demonstrated as widely as possible. Indeed, it is striking that so many mathematical methods that seem to be intimately connected with compressible flow are equally useful in areas ranging from solid mechanics to electromagnetism.

However, this is by no means the end of the story as far as the mathematical theory of wave propagation is concerned. In recent decades, there has been a spectacular blossoming of theory associated with traveling disturbances in chemical and biological systems as distinct from mechanical or electromagnetic ones. Such waves are described in detail by Lighthill [30] and especially by Billingham and King [17], where they are placed side by side with the waves we have discussed here. As is evident from these works, these "less classical" waves can exhibit many of the features that we have encountered in the preceding pages: steepening, dispersion, reflection, diffraction, and so forth. However, there is something formally distinctive about waves that are governed by systems whose linearization yields *real* dispersion relations between the wave number k and frequency ω. This, of course, includes the *high-frequency* behavior of every *hyperbolic* system. This means that these systems have a robustness in that they can exist without relying on any input or loss from their surroundings. On the other hand, when the wave is governed by a *parabolic* system, whose linearization can only yield a complex dispersion relation in which disturbances decay temporally, then its very existence requires a compensating steepening mechanism (usually via nonlinearity) and the wave propagates as a balance between the two. We have encountered this kind of delicate balance just once in this book, namely in Burgers' equation (6.11). However, without Burgers' equation, we would never have been led to our selection mechanism for shock waves, and the whole theory of nonlinear gasdynamics would be in ruins. The parabolicity in Burgers' equation comes about through the presence of viscosity, which is itself a result of molecular forces acting within the gas. Viscosity is a difficult concept that we have studiously tried to avoid in this book, so as not to expose the reader to the complexities of the compressible Navier–Stokes equations.

This is a familiar story in applied mathematics; although we may have hoped for a comprehensive self-contained theory of compressible flow based on macroscopic principles of conservation of mass, momentum, and energy, we have not been able to escape entirely from consideration of the intermolecular forces that determine not only the equation of state but also the correct macroscopic model for gasdynamics, especially in extreme configurations such as shock waves.

References

1. Ockendon, H. and Tayler, A.B. (1983) *Inviscid Fluid Flows*, Springer-Verlag, New York.
2. Chapman, S. and Cowling, T.G. (1952) *The Mathematical Theory of Nonuniform Gases*, Cambridge University Press, Cambridge.
3. Ockendon, H. and Ockendon, J.R. (1995) *Viscous Flow*, Cambridge University Press, Cambridge.
4. Glass, I.I. and Sislan, J.P. (1994) *Nonstationary Flows and Shock Waves*, Oxford Univesity Press, Oxford.
5. Acheson, D.J. (1990) *Elementary Fluid Dynamics*, Oxford University Press, Oxford.
6. Greenspan, H.P. (1968) *The Theory of Rotating Fluids*, Cambridge University Press, Cambridge.
7. Coulson, C.A. and Boyd, T.J.M. (1979) *Electricity*, Longman, London.
8. Love, A.E.H. (1952) *A Treatise on the Mathematical Theory of Elasticity*, 4th ed., Cambridge University Press, Cambridge.
9. Ockendon, J., Howison, S., Lacey, A. and Movchan, A. (1999) *Applied Partial Differential Equations*, Oxford University Press, Oxford.
10. Hinch, E.J. (1991) *Perturbation Methods*, Cambridge University Press, Cambridge.
11. Drazin, P.G. and Reid, W.H. (1981) *Hydrodynamic Stability*, Cambridge University Press, Cambridge.
12. Lighthill, M.J. (1978) *Waves in Fluids*, Cambridge University Press, Cambridge.
13. Courant, R. and Hilbert, D. (1962) *Methods of Mathematical Physics*, *Vol. I*, Interscience, New York.
14. Arscott, F.M. (1964) *Periodic Differential Equations. An Introduction to Mathieu, Lamé and Allied Functions*, Pergamon Press, Elmsford, NY.
15. Born, M. and Wolf, E. (1980) *Principles of Optics*, 6th ed., Pergamon Press, Elmsford, NY.
16. Chapman, S.J., Lawry, J.M.H., Ockendon, J.R., and Tew, R.H. (1999) SIAM Rev. **41**, 417–509.

17. Billingham, J. and King, A.C. (2000) *Wave Motion*, Cambridge University Press, Cambridge.
18. Russell, J.S. (1845) 143th Meeting, British Association for the Advancement of Science, York.
19. Kevorkian, J. and Cole, J.D. (1981) *Perturbation Methods in Applied Mathematics*, Springer-Verlag, New York.
20. Drazin, P.G. and Johnson, R.S. (1989) *Solitons: An Introduction*, Cambridge University Press, Cambridge.
21. Dodd, R.K., Eilbeck, J.C., Gibbon, J.D., and Morris, H.C. (1982) *Solitons and Nonlinear Wave Equations*, Academic Press, New York.
22. Garabedian, P.R. (1964) *Partial Differential Equations*, John Wiley & Sons, New York.
23. Whitham, G.B. (1974) *Linear and Nonlinear Waves*, John Wiley & Sons, New York.
24. Lax, P.D. (1953) Communs. Pure and Appl. Math. **6**, 231–258.
25. Liepmann, H.W. and Roshko, A. (1957) *Elements of Gas Dynamics*, John Wiley & Sons, New York.
26. Chapman, C.J. (2000) *High Speed Flow*, Cambridge University Press, Cambridge.
27. Courant, R. and Friedrichs, K.O. (1948) *Supersonic Flow and Shock Waves*, Interscience, New York.
28. Van Dyke, M. (1975) *Perturbation Methods in Fluid Dynamics*, Parabolic Press, Stanford, California.
29. Anderson, J.D. (1989) *Hypersonic and High Temperature Gas Dynamics*, McGraw-Hill Book Company, New York.
30. Lighthill, M.J. (1975) *Mathematical Biofluiddynamics*, SIAM, Philadelphia.

Index

Texts in Applied Mathematics